SUPER OPERATOR:

Practical Water Treatment Plant Operations and Training

BY. V.A. BROOKS

SUPER OPERATOR:

Practical Water Treatment Plant Operations and Training

ISBN: 978-0-8206-0413-8

First Edition:

Chemical Publishing Company:

www.chemical-publishing.com

Printed in the United States of America

TABLE OF CONTENTS

About the Author

Have you ever noticed that most "About the Author" sections sound like they were written by anyone except the author? This one is not like that.

My name is V. A. Brooks. I'm the Assistant Water Production Manager at the G. Robert House Jr. Water Treatment Facility in Suffolk, Virginia. I hold a State of Virginia, Class 1 Water Works Operator's license.

As a published writer, my focus has been in water treatment and other technical areas. I'm a strong advocate for the Water Treatment profession, I belong to the American Water Works Association (AWWA) Virginia Section. I have been a member of the Work for Water Committee whose primary focus is to bring awareness to the youth of Virginia about all areas of Public Utilities. I was also a member of the AWWA's Plant Operations Committee.

I attended Shortridge High School, Indianapolis, Indiana and Purdue University, West Lafayette, Indiana before entering the United States Navy. Several years after enlisting, I was selected for a commission and retired as a Surface Engineering Limited Duty Officer with twenty-one years of service. My background is in Main Propulsion Engineering, specifically Marine Gas Turbine propulsion with a specialty in power generation and electronics. During my naval career, I was also a hovercraft engineer and assistant craft master (pilot). You may be familiar with the U.S. Navy hovercraft known as the Landing Craft Air Cushion or LCAC (el-cack) for short.

Spending so much time working in Naval Engineering, I understand that knowledge of your profession makes you better at your job. Simple enough, right?

I believe you must have a firm foundation in the basics and build from there. This is why calculus is not your first math course.

I am a Super Operator, a trained professional, firmly grounded in the basics of water treatment. Armed with detailed knowledge of my plant and its systems, I am a force to be reckoned with. When I come to work, I bring a professional attitude and I'm focused on the task at hand.

Having traveled the world extensively and visiting many Third World countries, I've developed a passion for the water treatment profession. I'm excited about what I do. I invite you to join me in the ranks of truly superior water treatment professionals.

Super Operators!

Preface

In over 12 years of working in the water treatment industry, I have read a great number of books on water treatment. Many of these these books will help you pass licensing examinations, but they will not help you operate a water treatment plant. I have yet to find one that specifically addresses day-to-day operating. So, I thought that a book addressing real-world plant operations was needed. This is that book.

This book is written in plain English. You will find most chapters are only one or two pages. What you won't find is a lot of chemical and mathematical formulas. Nor will you find a bunch of incessant charts and graphs. Nothing wrong with formulas, charts, and graphs. There are lots of books in which you can find that.

Let me give you an example of what I mean. I have never found the chemical formula for aluminum sulfate, Al2(SO4)3 * 14 H2O to be useful when operating a water treatment plant. What is useful is knowing that aluminum sulfate is acidic in nature and tends to drive pH down. Now as an Operator, that's information I can use to better operate. This is what "Super Operator" is about—giving you practical information that will help you become a better Operator and run the plant more efficiently.

As you read the book you will notice that it sounds

like I'm talking directly to you. That's because you and I are having a conversation. And when we're done I want you to be a "Super Operator."

Directors, Managers, Supervisors, and Operators tell me that training Operators is an area of major concern. Ninety-five percent of all plant violations originate in the treatment process and should have been stopped by the Operator. Ninety-four percent of these violations can be traced to a lack of training. "Super Operator" is the first book to pointedly address water treatment as related to actual plant operations. Appendix A at the end of the book is a readymade, modifiable resource for training Operators on the plants in which they work.

Most plants don't have training programs for trainees and newly hired licensed Operators. By the way, assigning the new hire to a senior operator is not a training program. Appendix A lets you uniformly train Operators so that everyone is operating the plant in the same fashion using the same procedures. It also allows for documenting and certifying that the Operator is qualified to run the plant.

Appendix A is available in editable file form, so it can be tailored to your plant's specifics. Editability also lets you emphasize those areas of plant operations that you deem critical.

Many Operators are lacking the basics of conventional water treatment and the in-depth knowledge about their own plants necessary to be effective Operators. They have been shown what buttons to push, but don't understand why they are pushing them. When casualties arise, Operators find themselves calling the Plant Manager or Operations Superintendent to save them because they don't know the plant. I had an Operator call me at 11 o'clock at night, informing me of a failed chlorine pump and asking what they should do. My reply was, "You have six of them. Start another one."

"Super Operators" are grounded in water treatment fundamentals and have a firm grasp on how their plants operate. They distinguish themselves with consistently outstanding performance, surpassing their contemporaries by leaps and bounds.

I invite you to continue reading and I assure you that you will not be disappointed. Please come and join the elite ranks of the "Super Operator"!

Thank you!

V. A. Brooks

Super Operator

List of Figures

SECTION 1

Let's Get Started

Believe it or not, throughout the industry I've encountered what I call the "need to know mentality." It's an unstated policy that says you only need to know some things about the plant and not others. Some hoard information and wield this knowledge as a weapon, instead of sharing it to make the plant and its Operators stronger. If you subscribe to this train of thought, then you may not want to read this book.

I believe that every Operator has a need to know. When Operators are highly trained and know their plants well it only makes the team better. I continually learn about my plant, its equipment and processes. It's why I'm a "Super Operator."

Ladies and gentlemen, hold on tight. This book is different than any other book you've read on water treatment. The language is plain, direct, and to the point. I have tried to eliminate useless words and cut directly to heart of the matter.

This book will help you with the actual running of a water treatment plant. You'll find the chapters short and easy to read. The information is relevant and immediately useful. I'll tell you about real-world experiences. The plan is to make you a better Operator and hopefully a Super Operator!

Let's go!

Chapter 1

What This Book Is About

Making Operators—better Operators, Super Operators if you will—that's what this book is about. *Super Operator* gives you information that is immediately useful. Information that enhances your technical and professional knowledge.

This book takes an engineering approach to plant operations. An approach that says you need to know your plant in its entirety to be an effective Operator. After all, water treatment is a field of engineering, not unlike power generation, plastics manufacturing, or any other engineering discipline.

Super Operator is based on two core principles. First, you must have a firm grasp on the fundamentals of water treatment. Second, you must have intimate knowledge of your plant and how it operates. With these principles in mind, *Super Operator* solidly addresses actual plant operations.

Most books that concern themselves with training Water Treatment Plant Operators are about preparing you to pass a licensing exam. Unfortunately, these books don't help you when you're running a water treatment plant. These books are also not written by Operators. *Super Operator* is the first book written by an Operator for Operators.

Super Operator (*SO*) points out that Operators are licensed and board certified. We are state-sanctioned professionals, no different than lawyers, doctors, or Class A contractors. *SO* builds on the principles of conventional water treatment as applied to actual plant operations, giving you solid information in terms that are easy to understand.

SO focuses on the importance of systems and knowing your plant intimately. Through the understanding of systems, you'll see how an action in one part of the process can affect another part.

The book is laced with real-world examples of what Operators are likely to see in the plant. As you progress through the book, the goal is to make it easier for you to accurately interpret the data, to recognize problems and their causes.

In my own experience, and in speaking with Operators and Plant Managers, two things have become apparent: many Operators are lacking in the fundamentals of water treatment and plant operations, and many are unaware of how their plants work. These shortcomings act as hamstrings that hamper you when trying to assess, troubleshoot, and correct plant abnormalities. *Super Operator* is about helping to bridge these voids.

With many plant violations originating in the treatment process, training is a matter of paramount concern to Directors, Managers, Supervisors, and Operators. *SO* is a modifiable resource for the uniform training of Operators on the plants in which they work. Water treatment professionals agree that training Operators in a uniform fashion is difficult.

Super Operator hands the managers, supervisors, and operators a system that can be used to train, document, and certify Operators in the proper and professional operation of their plants. That system can be found in Appendix A.

Super Operators have mastered water treatment fundamentals. They know their plant intimately, especially the plant's systems and their relationships to one another. Super Operators understand how each action impacts the overall system down the line.

After reading this book, you should be stimulated to view the way you operate through the eyes of a professional, focused, certified Water Treatment Plant Operator. This is what the state says you are when they issue your license.

Super Operator is about being of immediate benefit to you, no matter if you're a veteran, junior, or entry-level Operator. Managers and supervisors will delight in having a training source that addresses the actual running of the treatment plant.

Ultimately this book is about making you a highly trained Super Operator!

Chapter 2

Water Treatment Plant Operators Are Professionals

Any Operator who has worked with or for me knows that I take my job very seriously. They understand that I expect them to know the plant, take corrective action when necessary, and bring their "A" game when they come to work. As an Operator, when everything is going south, this is when you make your money, not when everything is running great. I can teach a monkey to run the plant when everything is working fine. No offense to monkeys.

We must understand that treating water for human consumption is a serious matter. Our profession is not the performance of some mindless task. You hold the public's health in your hands. Your job demands a high degree of intelligence, knowledge, pride, and professionalism, and is based on a great deal of trust.

As Water Treatment Plant Operators, we sometimes fail to realize we're professionals. We don't fully grasp the magnitude of our responsibility. Nor do we see the dire consequences that can happen if we fail in the execution of our duties.

Operators are certified by the state and required to maintain a certain amount of continuing education.1 That means you are no different than a doctor or lawyer. When you made the decision to become licensed, certified and authorized by the state, it was a career decision.

Think about this. The entire state of Virginia has 734 Class 1 Operators.1 Put another way, only nine-thousandths of a percent (0.009%) of the 8.4 million people in the state of Virginia 2 hold a Class 1 Waterworks Operator license. This means you belong to an elite group of trusted agents. There are very few authorized and able to do what you do.

Water Treatment Plant Operators are professionals, and we have an awesome responsibility. Every time you assume the duties of a shift Operator you are holding the health and wellbeing of numerous lives in your hand. In the case of larger plants, you may be supplying water to hundreds of thousands of customers.

Shift Operators are the first line of defense for the citizens of their communities. You are responsible for keeping various bacteria, cysts, waterborne diseases, and other detrimental pathogens out of the distribution system.

Operators are required to know a multitude of water quality standards set forth by federal, state, local, and plant-specific organizations. Your job, "Water Treatment Plant Operator" is not a game. It takes focus and a professional attitude to be an effective Operator. Ultimately this profession demands your best.

Anyone who's been in water treatment for a while has heard the phrase, Operator In Responsible Charge (OIRC). Many Operators are designated knowingly and unknowingly as "Operator In Responsible Charge." What does this really mean?

"Operator In Responsible Charge" (OIRC) is a designation placed upon the Operator by the Owner of the water treatment facility. This designation is so important that most state administrative codes mandate the assignment of an Operator who is in charge or accountable for the treatment process. These codes go into exacting detail regarding who can be designated as the OIRC.2,3

The OIRC is responsible for the control of plant operations, testing, and evaluation of the process. The OIRC must ensure the finished water meets various federal, state, local, and plant regulations, standards, and benchmarks.3,4

The OIRC is regulated by such agencies as the Environmental Protection Agency (EPA), the Department of Environmental Quality (DEQ), the Department of Health (DH), the Occupational Safety and Health Administration (OSHA), and a host of other alphabets with deep regulatory teeth.

These agencies have the power to regulate, and these regulations have the force of law. So, my fellow Operators, be advised that your performance or lack thereof can and often does have legal ramifications. Please consider the following Real World 101 cases:

Real World 101

Case #1 – New Brunswick, Jersey (2015)

Chief Operator of the New Brunswick Water Utility plead guilty to charges of public corruption and violations of the Safe Drinking Water Act violations. Prosecutors are asking for a three-year sentence.5

Case #2 – Dolton, Illinois (2015)

An Operator in Dolton, Illinois is charged by the Department of Justice, U.S. Attorney's Office, Northern

District of Illinois with falsifying sample results, violating the SDWA and various EPA regulations.6,7

Case #3 – Sebring, Ohio (2016)

The Operator of Record of a public water system in a village near Youngstown has been charged with failing to give timely notice to homeowners and residents. After receiving lab results with high lead levels, Ohio safe drinking water laws require that affected individuals be notified within 30 days.8

Case #4 – Steelton, Pennsylvania (2015)

A Steelton Water Treatment Plant Operator is facing criminal charges by the state Attorney General's Office. The Operator allegedly falsified records related to the safety of local drinking water. Tampering with public records or information is a clear violation of the Safe Drinking Water Act.9

I can go all day with this and I haven't even touched on Flint, Michigan. My hope is that these real cases will open your eyes to the serious nature of our profession. I hope this chapter has shown you that Water Treatment Plant Operators are professionals.

REFERENCES:

1. Alexina M. Borkey, Regulant Lists, Information Management, Department of Professional and Occupational Regulation (DPOR), Virginia Records Management, publicrecords@dpor.virginia.gov, Richmond, VA, November 21, 2016.

2. United States Census Bureau, Quick Facts Virginia, census.gov/quickfacts/table/PST045215/51, 01 JUL 2015.

3. Virginia General Assembly, Water Works Regulations, Virginia General Assembly, Richmond, VA, 21 JUN 2012.

4. Colorado Department of Public Health and Environment (CDPHE), Operator in Responsible Charge (ORC), CDPHE, colorado.gov/pacific/cdphe/wq-operator-responsible-charge-orc, Denver, CO, 2016.

5. Charlie Kratovil, NB Water Official Finally Resigns and Pleads Guilty to Corruption, NewBrunswickToday.com, New Brunswick, NJ, 17 DEC 2015.

6. Jordan Owen, Former Dolton Water Operator Charged with Falsifying Water Sampling Data, chicago.suntimes.com, Chicago, IL, 11 JUN 2015.

7. Department of Justice, U.S. Attorney's Office Northern District of Illinois, Former Dolton Certified Water Operator Charged with Falsifying Drinking Water Sampling Data, justice.gov, Chicago, IL, 11 JUN 2015.

8. Associated Press, Ex-water Operator Charged over Lead in Ohio Village's Taps, foxnews.com, Sebring, OH, 13 JUL 2016.

9. Megan Trimble, Former Steelton Water Chief Operator Charged in Public Record Tampering, pennlive.com, Steelton, PA, JUN 2015.

Chapter 3

Systems and Their Importance

Now it's time to get to the heart of the matter. Simply put, you need to have a thorough understanding of systems to really know your plant.

"System" as defined by the *American Heritage Dictionary*: "A group of interacting, interrelated or interdependent elements forming a complex whole . . ."1

A water treatment plant is one big system that has many component systems. Those component systems in turn interact and work together to produce quality drinking water for the public. Common systems include raw water, chemical storage and delivery, process water (finished water that is used in-plant to treat raw water, usually labeled nonpotable), residuals monitoring, and supervisory control and data acquisition (SCADA) systems—and that's just the tip of the iceberg.

A Super Operator must know how they operate and how they are related to other systems. When you begin to understand each system and its relationship to other systems, you can see how an action in one system affects another system down the road. You need to know equipment types, their operational modes, flow capacities, speed, etc. What inputs make them function? For example, most chemical feed pumps receive flow and dosage inputs; from these inputs the pumps calculate the speed (and stroke if applicable) needed to achieve the desired dosage.

High-service pumps that supply finished water to the distribution system are another example. These pumps usually have automatic, manual, and local operating modes. In automatic they are typically designed to run in pressure or flow mode. Pressure mode maintains a desired pressure set point and flow mode ensures the pump delivers a set amount of flow independent of system pressure. Operators need to know what these modes mean and how the pump should behave in each mode.

Let's see why knowledge of system interrelationships is important. We'll stay with the high-service pump system. The next four paragraphs describe a real scenario that should reinforce the importance of having in-depth knowledge of systems.

Let's say the high-service pump system is operating in pressure-regulating mode. The system is set to maintain 68 pounds per square inch (psi). The following information is critical to successfully correct the casualty.

You need to know that in pressure-regulating mode the high-service pump system relies on input from the sensor that reads discharge pressure. This sensors location and where its output is displayed as well as the location of other pressure sensors that monitor the same parameter, all of which will be vitally important to you.

On the control room SCADA screen you receive a low high-service discharge pressure alarm and notice your high-service pump is running at 100 percent. The discharge pressure reading is zero (0) psi.

An Operator that knows the plant's systems realizes that in automatic pressure-regulating mode the pump sees zero psi and speeds up, trying to maintain 68 psi system pressure.

Having a thorough knowledge of systems, you go to the high-service pump room and read an analog pressure gauge that reads the pump discharge pressure as 85 psi. Knowing your pumps are variable speed, you place the pump into manual control and regulate the speed until you get the desired 68 psi on the functioning analog gauge.

In-depth knowledge of systems is a characteristic common to all Super Operators. Knowing the plant's systems will help you troubleshoot problems. It enables you to interpret normal and abnormal plant responses and behavior. This knowledge will help you improve plant performance and quicken operational responses to malfunctions.

Like the above scenario, below are a couple of real-world situations. Throughout the book, you'll come across REAL WORLD 101 sections that tell of real situations and other useful information. These case studies should help to illustrate the critical importance of understanding how your plant systems work and interact.

Real World 101:

Case #1:

Your ammonia and monochloramine analyzer has been reading zero ammonia residual for a couple of hours. The free chlorine residual is stable. Monochloramine and total chlorine residual are beginning to drop. A thorough understanding of your ammonia/monochloramine analyzer will tell you that you have an ammonia feed problem, not a chlorine problem. The zero-ammonia residual for an extended time should have been your first clue. Unfortunately, most Operators see the total chlorine residual dropping and increase the chlorine feed, which makes the problem worse. The total chlorine and monochloramine are dropping because it's turning into free chlorine due to insufficient ammonia being fed. The last thing you want to do is add more free chlorine.

Case #2:

The #1 raw pump is set to maintain a flow of 10 million gallons per day (MGD). You notice the flow is fluctuating radically between four and 12 MGD. It's two o'clock in the morning and maintenance can't get there for three hours. Because you know the system, you switch to another raw pump and see that your flow remains erratic. Placing the pump in manual and trying to control the flow through the SCADA system does not work either. You decide to go to the pump's local controller, place it in local, and regulate the flow manually. Maintenance arrives in the morning and finds a bad control card in the Programmable Logic Controller (PLC).

In the above real-world scenarios, your knowledge of the systems stopped you from turning a chloraminated system into a free chlorine system in case #1, and you averted a possible plant shutdown in case #2.

REFERENCES:

1 *American Heritage Dictionary*, 5th edition, Houghton Mifflin Harcourt. 2011, Boston/New York, p. 1768.

Chapter 4

Why You Must Know Your Plant Intimately

Chapter 3 discussed systems and their importance. This chapter expounds upon chapter 3 and ties in systems as we look at the plant in its entirety. A thorough understanding of your plant and how it works is integral to your ability to effectively operate the plant.

The Operator who doesn't possess the requisite knowledge disadvantages the treatment team. In fact, every job description I've read on Water Treatment Operators states you must have that knowledge to properly perform the essential job functions. Extensive understanding of the plant and how it operates is mandatory.

I've been on shift with Operators that were alarmingly unaware of how the plant worked. They were sorely lacking in the fundamentals of water treatment. These Operators needed to be monitored so closely, that I was in effect running the plant alone. On the other hand, if your fellow Operator knows the plant and is proficient in its operation, your shift will run much more smoothly.

To run the plant efficiently you have to know how it works. If you don't, when problems arise you won't be able to make proper decisions. With the ever-automated functions of the modern water treatment plant, in-depth knowledge of how the equipment works is necessary to determine malfunctions, make corrections, and interpret data. Knowing the plant intimately enhances your ability to make timely, accurate assessments about the treatment process.

If you want to distinguish yourself from average operators, thorough knowledge of how your plant works is essential. Most plant casualties happen after hours when everyone has left the plant. At one o'clock in the morning when turbidity is rising and the plant is coming apart, you don't have the time or the luxury of waiting for maintenance or the Water Production Manager to come and save you.

I tell my operators that when things are going south is when they need to step up and get the plant back on track. These are the times for which you really get paid. These are the times that the government (federal and state) and your community look for you to keep them safe as professionals and as the Operator In Responsible Charge (OIRC).

Your intimate plant knowledge will assist you when assessing situations and troubleshooting problems. Knowing the plant is your foundation for achieving peak plant performance. It also helps you avert problems in treatment before they occur and prevent plant violations.

You can see the enormous benefit that's afforded when you thoroughly understand how your plant works. This knowledge is crucial for you to elevate yourself to the level of Super Operator. The Real World 101 scenarios below should make this even clearer.

Real World 101:

Case #1:

Your plant has multiple clarifiers with two normally online. The rapid mix has separate trains but feeds the clarifiers from a common header. Alum is the primary coagulant and is fed into individual rapid mix trains and then discharged into a common header. Polymer is fed to each clarifier from a dedicated polymer pump.

You notice that clarifier 2 turbidity has risen gradually from the 0.400 range to over 1.00 Nephelometric Turbidity Units (NTU). Clarifier #1 turbidity remains stable at approximately 0.400 NTU.

The Super Operator sees that only clarifier 2 has been affected and that the most likely culprit is the polymer feed to clarifier 2. An alum problem would have affected both clarifiers because they are from a common header after the rapid mix. A quick drawdown of the polymer pump feeding clarifier 2 shows that it's pumping less than half of the set dosage. You have four polymer pumps that are cross-connectable. Polymer pump 2 is stopped, and polymer pump 3 is aligned to feed clarifier 2. Once polymer pump 3 is started, a drawdown is done, which confirms the pump is feeding the desired dosage. Clarifier 2 turbidity begins to drop within 20 minutes.

Case #2:

During normal plant operations, you lose carbon feed to one of your online sedimentation basins. Each basin has one carbon pump assigned to it. Because you have expert plant knowledge, you know that you can use an offline carbon pump by connecting jumper hoses for just such emergencies until maintenance can take a look at the damaged pump.

Chapter 5

You’re a Drop of Water

When I really want to know whether an Operator knows the plant, I have an exercise I ask them to do. It’s called “You’re a drop of water...”

YOU’RE A DROP OF WATER

TAKE ME FROM THE SOURCE

TO THE CLEAR WELL . . .

"You're a drop of water" works like this. I say to the Operator, "You're a drop of water. Take me from the source to the clear well." I tell Operators that I want to know every step the water drop goes through to get there. The Operator should start in the source water. This could be a lake, river, reservoir, etc. From the raw water source, you (the water drop) might get drawn into a raw water pump. From there your next destination might be a rapid mix area where the primary coagulant is added. Feel free to stop them as they go along or wait until they are done. I usually wait when the Operator is giving me a lot of detail and question intermittently when they are vague and trying to skip through and get the exercise over with quickly.

For example, the exercise may start out like so:

V. A. – "Hey, Alisha, you're a drop of water. Take me from the raw source to the clear well, and tell me in detail everything that happens along the way."

Alisha – "Okay, I'm in Lake XYZ and get drawn into the intake and through the raw pump. Then . . ."

V. A. – "Hang on a minute! Tell me about the raw pumps. What kind of pumps are they? Positive displacement? Centrifugal? How many are there? What are their rated capacities? What are their modes of operation? Auto? Manual?"

You may want to start out by allowing the Operator to describe the basic flow path in which all the major points of reference are mentioned. As the Operator increases in knowledge, then more intensive questions should be asked.

And the exercise proceeds thusly. As you're going through the process you should be able to tell the story of the water drop as it moves from source to clear well, explaining each step of your journey as you become a clean, safe, and delicious droplet of fresh drinking water.

"You're a Drop of Water" gets you to envision the overall process. It puts a picture in your mind of all the systematic goings on. It's a priceless tool, for it helps you learn the plant.

Hopefully you can see that this exercise will help you as an Operator. It will give you a better grasp of your plant's overall process. When things are coming apart you'll be able to quickly isolate where the problem is and implement sound corrective actions. All because you were a drop of water.

All my Operators are painfully familiar with the "You're a Drop of Water" exercise. They'll say that they are better Operators because of it.

SECTION 2

Chemicals 101

Conventional water treatment involves chemical and mechanical processes. Both processes are very important and can cause your treatment to go south very quickly if they are not implemented and functioning properly.

It's important to know what the various treatment chemicals are used for, as well as their characteristics. The following chapters in this section will describe the basic makeup, various functions, applications, and characteristics of the most commonly used chemicals. Formulas will be pretty much nonexistent. As a Super Operator I have yet to find an instance where knowing that the formula for liquid alum—$Al_2(SO_4)_3$—was actually useful.

What is useful to know is that liquid alum is acidic and characteristically drives pH down.

Real World 101:

On shift I notice that my applied pH is dropping rapidly. Since I know that alum is acidic my problem almost assuredly lies with my alum feed. I take a look at my chemical pump control page for alum and I see that the pump has inadvertently been placed in manual instead of auto. The manual setting has a pump speed of 67 percent, which is too much for the established raw flow and set dosage. This explains the plummeting pH.

Regarding the above incident, I asked a junior Operator what could cause such a drastic drop in pH. When he answered saying "a carbon overfeed," I was flabbergasted. A carbon overfeed could not possibly have caused the pH to drop. In fact, too much carbon would have driven the pH higher, and the rise in pH would have been negligible.

This is just one instance where understanding the chemicals you're working with makes you a better Operator.

I hope that you can see a recurring theme!

Chapter 6

Primary Coagulants

If you're going to talk about conventional water treatment chemicals, you should start with primary coagulants. Yes, sometimes pretreatment chemicals are added prior to—but without coagulation, ladies and gentlemen, you've got nothing!

What is coagulation and why do we need it? In water treatment when we speak of coagulation we're talking about the process that gets small unwanted particles in water to combine and settle out in a reasonable timeframe. These small particles are often referred to as "nonsettleable" solids or colloids. Colloids are defined as "finely divided particles, which are approximately 1 to 1,000 millimicrons in size, are dispersed within a continuous medium in a manner that prevents them from being filtered easily or settled rapidly."1

Okay, so we're saying that raw water contains these unwanted tiny solid particles that are hard to get out. Whew!

It's fairly easy to get larger solid particles to settle out through the sedimentation process, but these smaller nonsettleable solids require a little help to get them to settle out in a timely manner. Enter the coagulation process. Scientists long ago figured out that we can make this happen by adding chemical compounds to the water. We call these compounds "Primary Coagulants."2

When we add these primary coagulants, they cause the solids in the raw water to come together and form larger particles. These larger particles are much easier to remove through natural settling and filtration. This forming of larger particles is called flocculation.2 The larger particles are referred to as "floc."

Bear with me as I get slightly technical. Raw surface water typically carries a negative charge. The addition of these primary coagulants, which are usually positively charged, facilitate the formation of floc. This is the reason we use primary coagulants like alum. Formally called aluminum sulfate. Basic physics tells us that opposites attract each other. By chemical addition, combined with rapid mixing and slower physical agitation, we can make the flocculation process begin.2

Note that primary coagulants are usually acidic and have pH's on the low end of the scale. It's their positive charge that makes them acidic. Remember the acidic nature of these primary coagulants. This is very useful knowledge to you as an Operator.

Since alum is acidic, when you add more, it tends to drive the pH down; less of it causes the pH to rise. Having read more than 50 plus books on water treatment, I have yet to find one that conveys this simple yet priceless fact in plain language.

Real World 101:

Your plant regulates its optimum (best) operating pH (also called applied pH) through its alum feed. You notice that your applied pH is rising, even though the SCADA is trying to maintain the pH automatically. As a Super Operator, you know this because your SCADA display shows you that the alum pump is being commanded to run faster, so that more alum will be fed. Knowing alum is acidic and drives pH down, you go to the pump and perform a drawdown. The drawdown shows that the pump isn't pumping efficiently. In fact, it is grossly underfeeding. Which explains the rise in pH.

Several common coagulants are listed below, but we're going to be speaking specifically about the most common, which is alum. The other coagulants in Table 6-1 have similarly characteristic behavior.

TABLE 6-1 COMMON COAGULANTS

COMMON NAME(S)	CHEMICALS NAME(S)	FORMULA
Alum, Liquid Alum	Aluminum Sulfate	$Al_2(SO_4)_3$
Ferric, Ferric Chloride	Iron (III) Chloride	$FeCl_3$
ACH	Aluminum Chlorohydrate	$Al_2Cl(OH)_5$
PACl	2 Poly Aluminum Chloride	$Al_n(OH)_mCl_{3n-m}$

Alum is overwhelmingly the primary coagulant of choice used in conventional water treatment. The reason it's so popular is because it's effective, low cost, and easy to store and handle.

Alum has an amber color and should not be cloudy; rather it should be clear and shiny. If it's in a transparent container, you would be able to see through it. Most alum loads have been recently mixed and are usually very warm to downright hot. I have taken loads in excess of 120 degrees Fahrenheit.

Many plants require a sample of the alum they receive before accepting delivery. We'll talk about receiving chemical loads in chapter 30. If you are required to take a sample, you will most likely be ascertaining the specific gravity with a Baume hydrometer.

Alum works best in a general operating pH range around 5.4 to 7.4. Above or below this range I find that you don't get coagulation. Without coagulation, you might as well be feeding raw water. Also, note that your primary coagulant will have an optimal operating pH that is much more exact and yields the best turbidity. Some factors that determine this optimal pH are season, temperature, types of pretreatment, and the raw water turbidity to name a few.

Over- and underdosing can cause high turbidity and in extreme cases the coagulation process will not occur. Overdosing can also cause an undesirable effect in the clear well, known as reflocking. Too much alum in the finished water can cause floc formation to start in your clear well. Not good. Reflocking causes high turbidity (cloudiness) in the customer's finished water, resulting in water quality complaints. These complaints translate into headaches for you.

Real World 101:

In order to cause the coagulation and flocculation processes to start, you must have alkalinity in the source (raw) water. So if you're troubleshooting a high-turbidity situation and you have zero alkalinity, check your alum feed and applied pH. The pH is probably too low, indicating an alum overfeed. Your alum has eaten all the alkalinity. A general rule of thumb is that it takes about two and a half (2.5) parts alum to consume one (1) part alkalinity.

References:

1. *American Heritage Dictionary*, 5th edition, Houghton Mifflin Harcourt, Boston/New York, 2011, p. 363.

2. *WSO: Water Treatment*, 4th edition, Nicholas G. Pizzi, American Water Works Association, 2010, pp. 47–49.

Chapter 7

Powdered Activated Carbon

Powdered Activated Carbon (PAC) is primarily used to control taste and odor problems. It can also be used for algae control.[1] For our discussion we will only be addressing its use for managing taste and odor issues.

Powdered Activated Carbon (PAC) is carbon that has been put into powdered form after it has been activated. You can make activated carbon from various sources such as peat, petroleum residues, wood, coal, and even nutshells. A major source used in water treatment is coal. This coal is heated in an atmosphere without oxygen so it won't actually burn, a process that changes the coal into carbon. The "activation" occurs when it is exposed to a blending of air and steam. A crushing and screening process then takes place before packaging.[1]

PAC is most commonly shipped in various dry quantities, from rail cars, large container trucks (18-wheelers), to sacks ranging from 800 pounds down to 50 pounds. The PAC is usually turned into a slurry by adding water and mixing at the treatment plant. Depending on plant size, this mixing can take place in large holding tanks with capacities of tens of thousands of gallons or small mixing tanks that hold only 50 or 100 gallons.

Powdered Activated Carbon is very dusty and the dust is super fine and light, easily airborne. It's hard to keep confined and will find a way to get out. If you walk into a super clean, pristine carbon room, the system is either shut down or inoperative. Either way it's not feeding carbon.

Overfeeding carbon can have a negligible effect on pH, causing it to rise slightly. It is more likely that you will see an increase in turbidity. In other words, significant PAC dosage increases will increase turbidity, and significant dosage reductions will lower turbidity. That's good to know.

A typical point of application would be prior to the settling or clarification process. I've seen it added with the rapid mix and just after the rapid mix. Figure 7-1 demonstrates a rudimentary carbon feed system.

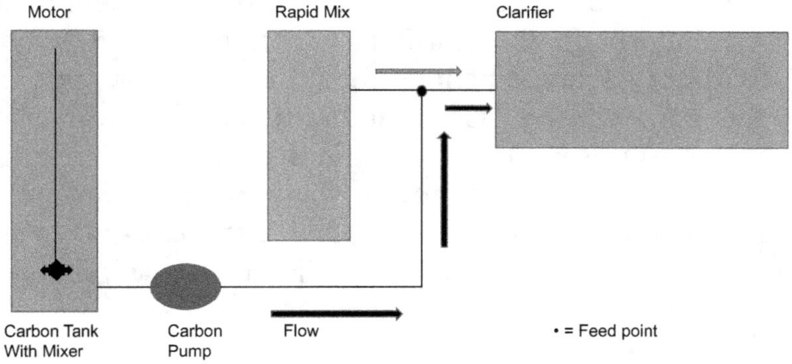

FIG 7-1 BASIC CARBON FEED SYSTEM

Carbon has a natural ability to remove taste- and odor-causing organics, both natural and synthetic. These organics are almost always found in surface water. It removes taste and odor through a process called adsorption.

Adsorption is defined as "The accumulation of gases, liquids, or solutes on the surface of a solid or liquid."[2] Adsorption, simply put, is the process by which materials are made to adhere to one another. Make no mistake, this is a complex physical and chemical process, but basically the activated carbon acts like glue and gets the unwanted taste- and odor-causing organics to stick to it.[1]

Your customers will make you aware of taste and odor problems. Ideally you want to keep your process running in a way that taste and odor does not become a problem for your customers. There is a test that checks the smell threshold odor of your water and assigns it a number, surprisingly called the Threshold Odor Number (TON), a test that is done daily in most water treatment plants. The TON test is usually performed by the Water Quality Technicians in the lab.

Odor levels can become so extreme that the smell can be quite noticeable throughout the treatment plant, particularly in the clarifiers or sedimentation basins. Without adequate actions for taste and odor removal, this odor will be passed on and remain in the finished water. At this point you'll begin to receive water quality complaints from customers. Many plants are aware of seasonal algal blooms and other taste/odor-increasing events and can handle these incidences without too much trouble.

Real World 101:

You notice a slight reduction in the sedimentation basin or clarifier turbidity—perhaps as low as 0.200 Nephelometric Turbidity Units (NTU) below what you have been recording. For example, if your clarifier normally runs about 0.600 NTU, and for the last four hours it's been running around 0.400 NTU, and the reduction is unexplained, check your carbon feed.

Turbidity is a measure of the cloudiness of water. It is commonly measured in Nephelometric Turbidity Units (NTU). We'll go in depth regarding turbidity in chapter 17.

References:

1. *WSO: Water Treatment*, 4th edition, Nicholas G. Pizzi, American Water Works Association, 2010, pp. 18, 343–45.

2. *American Heritage Dictionary*, 5th edition, Houghton Mifflin Harcourt, Boston/New York, 2011, p. 24.

Chapter 8

Polymers

Polymers are typically used to help the treatment process in doing a better job of removing unwanted material in the water being treated. A polymer is defined as "Any of numerous natural and synthetic compounds of usually high molecular weight consisting of up to millions of repeated linked units, each a relatively light and simple molecule."[1]

Polymers are sometimes referred to as polyelectrolytes and come in powder and liquid form. In powdered form a mixing system will be required.[2] Most plants buy powdered polymers and mix on site. Why use powder? Follow the money: using powder is more cost effective, even when you factor in the cost and maintenance of having to employ a special mixing system.

Normally polymers will be employed as a coagulant aid or an aid to filtration. How it is used is determined by the point of application. When fed just after the primary coagulant and the rapid mix, the polymer is being used as a coagulant aid. If applied just before the filter, it's being used as an aid to filtration. Makes sense, right? Polymers can also be used as primary coagulants but the cost would be prohibitive.

Polymers come in three forms: Anionic, Cationic, and Nonionic.

Anionic polymers: Produce ions that are negatively charged when dissolved in water.

Cationic polymers: Produce ions that are positively charged when dissolved in water.

Nonionic polymers: Are neutrally charged when dissolved in water. Understand that nonionic polymers in fact release both positively and negatively charged ions that result in a neutral solution.

These three categories of polymers all to some degree help reduce turbidity as well as provide for better settling and floc formation.

When feeding polymer, under- and overdosing (feeding) will both result in high turbidity. Like primary coagulants, your polymer feed will have an optimum range depending on various factors such as temperature, season, raw turbidity, etc.

BASIC MIXING SYSTEM

Figure 8-1 depicts a basic manually operated feed system. It's comprised of two feed tanks with mixers. An associated drawdown tube is installed to verify actual pumping rates. A flushing system for cleaning the lines when pumps are changed is seen against the wall. The polymer feed tanks are manually batched by the Operator. The feed tank is filled with water to the required level. Then the Operator turns on the mixer and slowly adds a premeasured amount of polymer. The polymer is added slowly to prevent clumping in the feed tank. Polymer batches are usually mixed for a set amount of time—30 minutes in this case. Once the mixing time has elapsed, the Operator stops the mixer and the batch is ready for use when necessary.

Polymer batches have a shelf life once mixed. This type is good for 24 hours. After the batch's shelf life has expired, the polymer chains break down and it's no longer effective in aiding coagulation or filtration.

MANUALLY OPERATED POLYMER PUMPS

Figure 8-2 shows manually operated polymer pumps. Operators set the pumps to their correct dosage by using the raw flow in millions of gallons per day (MGD), drawdown quantity, and time. From these three pieces of information the exact dosage can be ascertained.

FULLY AUTOMATED POLYMER SYSTEMS

Figure 8-3 shows an automated system that is typical of what you might find in a modern plant. To the far right the box-like container is the hopper that contains the dry polymer powder. It is equipped with a screw feeder, hot air blower, low-level sensor, and an automatic exhaust fan for dust suppression when opened. This unit is also capable of feeding liquid polymer and has an associated drawdown tube as well.

In the center is the brain of the system. On the control panel is a human/machine interface (HMI) touchscreen with manual control knobs below. The system can be run in automatic and manual from the touchscreen with all set points and safety limits enabled, as well as manually, using the control knobs with set points and safety limits disabled. The display shows the process in real time as it goes through each step. It can be stopped at any point and picked up where it left off with the touch of a button. A last-batch function provides for disabling the automatic rebatching function when required. There are two large tanks behind the control panel; the uppermost tank is the batch tank and the lower tank is the feed tank.

The batch tank has a polymer mixing head that blends the dry polymer and water as it enters the tank. A mixer starts automatically when the desired strength is met, and mixes for 30 minutes. Once the feed tank gets low, the polymer in the batch tank will be released into the feed tank and a new batch of polymer is made.

Against the wall to the far left are two redundant polymer feed pumps with associated local control panels. These pumps located below the control panels run automatically, feeding the set dosage. Of course, they can be operated manually if necessary. As with all chemical feed systems, a drawdown tube is installed to verify actual pump performance.

Figures 8-4, 8-5, and 8-6 below display the dry hopper, HMI operating panel, and redundant polymer feed pumps respectively.

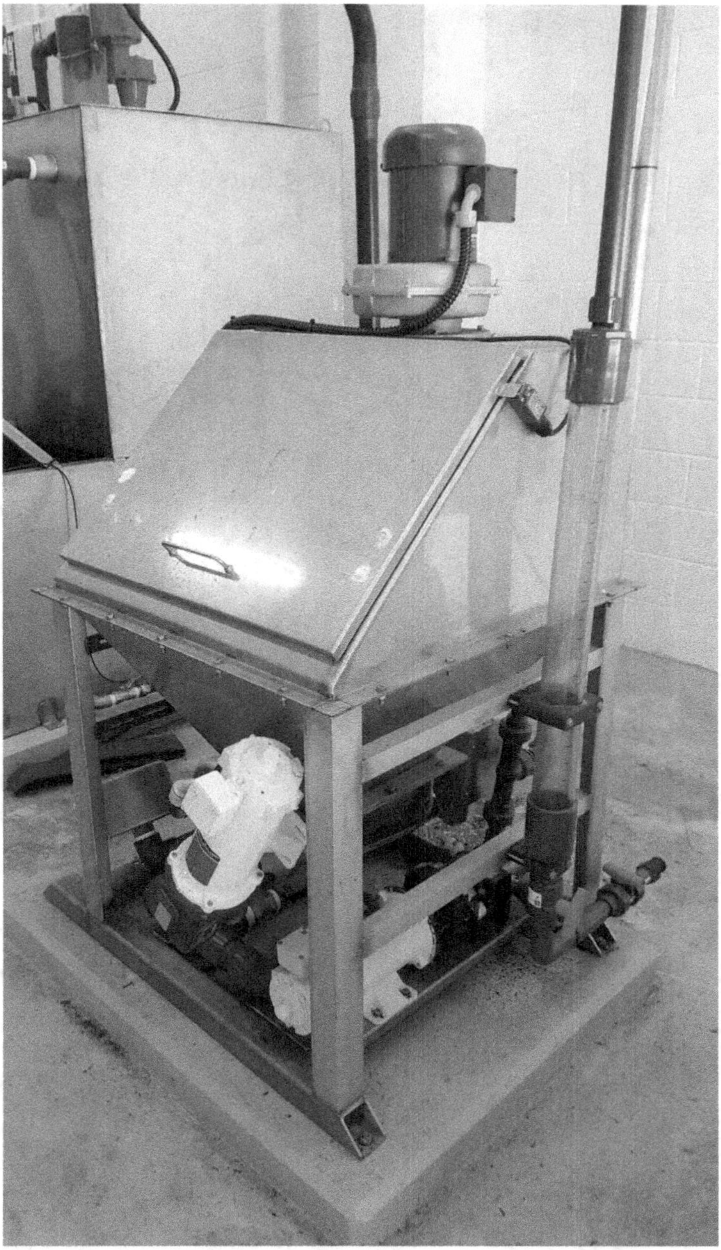

DRY POLYMER STORAGE SYSTEM

HMI CONTROL PANEL

AUTOMATIC POLYMER FEED PUMPS

Real World 101:

When polymers are hydrated (meaning water has been added) they will have a mucus-like consistency.

Polymer is usually made in batches with automated or at least semi-automated mixing systems. The typical system will alternate between feed tanks. Once the online tank transmits a low-level signal, the system will automatically shift to the standby tank. The system will then automatically rebatch the offline tank.

Occasionally the mixing system may produce a weak batch of polymer. A weak batch can have many causes. Some common causes are:

a. Improper feed of the dry polymer.
b. Improper water flow rate or low pressure.
c. Mixer failure.
d. Low levels of dry polymer in the feed hopper.

In any event this is bad. A weak batch of polymer is tantamount to a drastic dosage reduction. Feeding a low-strength (weak) batch of polymer will result in high turbidity. If you notice high turbidity after a shift to a different feed tank, check your polymer batch.

Since properly mixed polymer has a thick mucus-like consistency, a weak batch will feel watery and thin to the touch. A check can be made by putting on rubber gloves and checking the polymer's consistency.

Should a weak batch be the cause of your rising turbidity, a thorough inspection of your mixing system should be done. Usually batching problems can be corrected quickly and a new properly mixed batch placed online. Most plants are equipped with redundant batching systems.

References:

1. *American Heritage Dictionary*, 5th edition, Houghton Mifflin Harcourt, Boston/New York, 2011, p. 1366.

2. *WSO: Water Treatment*, 4th edition, Nicholas G. Pizzi, American Water Works Association, 2010, pp. 55, 135.

Chapter 9

Oxidizers

Oxidizers are used to remove or reduce undesirable elements in source water through a process called oxidation. For this discussion we will focus our attention on two troublesome elements, iron and manganese. Iron and manganese are common elements that can be found in excessive levels in surface water. They spawn aesthetic problems in drinking water. High levels of these elements when contacted by air will cause discoloration and impart odors that make the water unpalatable to the consumer.[1] They're excellent food sources for bacteria. Left unfettered, these bacteria can clog pipes, reduce flow, discolor water, and increase chlorine demand.

Through the oxidation process we can get soluble (dissolvable) forms of iron and manganese to precipitate out as insoluble compounds.[1] Precipitate here means to cause a substance to be deposited in solid form from a solution. Manganese and iron will come out as manganese hydroxide and ferric hydroxide or oxyhydroxides. The precipitate can then be settled or filtered out.

Oxidation is a complex chemical process. I won't get into the technical nitty gritty of it. Succinctly, oxidation always involves the transfer of electrons and protons. One substance gives up and one substance accepts, always leaving balance.

What you need to know is that by using oxidizers we can remove undesirable elements from the water we're trying to treat.

Table 9-1 below lists chemicals used to oxidize undesirable elements from raw water.

TABLE 9-1 COMMON OXIDIZERS

COMMON NAME(S)	REMARKS
Potassium Permanganate	Purple powder, stains on contact. Stains are hard to remove. Cost effective, no known trihalomethane (THM) formation.[1]
Sodium Permanganate	Purple liquid, stains on contact, hard to remove, leak-prone. Cost effective, no known trihalomethane (THM) formation.[1]
Sodium Hypochlorite	Bleach, cost effective, easy to store. Disinfection by-products and THM precursors are an important concern.
Ozone	Extremely powerful oxidant. Expensive and very corrosive.[1]
Air	Can be effective in oxidizing iron.[1]
Chlorine Dioxide	Powerful oxidant, highly water soluble, formation of by-products chlorate and chlorite are a matter of concern.[1]

Potassium permanganate powder is typically delivered in 55-gallon drums and requires a feeder system of some type. Screw feeders in conjunction with mixing water are often used to accomplish this. Sodium hypochlorite and sodium permanganate are liquid and come in various container sizes; large plants will have it delivered in bulk on large trucks. Chlorine dioxide is usually generated on site. Most plants are moving away from its use due to its volatility and propensity for chlorinated by-products, chlorite and chlorate formation. Ozone can't be stored, so it must be generated on site. It has a short life and is not very water soluble (capable of being dissolved). Ozone off-gas must be collected from the contactor because these off-gases are toxic. Its use is heavily regulated in some areas of the United States.

Oxidizers are usually fed early in the treatment process towards the head of the plant. They have an inverse relationship with turbidity and chlorine demand.

Real World 101:

If you lose feed on your oxidizer you will find that both applied turbidity and chlorine demand will rise.

References:

1. *WSO: Water Treatment*, 4th edition, Nicholas G. Pizzi, American Water Works Association, 2010, pp. 194–97, 277, 279–81.

Chapter 10

Chlorine

Chlorine can be used in oxidation, taste and odor control, algae and slime control, and pretreatment processes. Having said that, it is overwhelmingly used for disinfection. Water plants use chlorine gas, calcium hypochlorite, or sodium hypochlorite for disinfection.

Most plants are getting away from using chlorine gas because it's dangerous and volatile. Chlorine gas is difficult to store and handle. It's highly prone to leaks and will quickly overcome an operator as concentration builds. When using chlorine gas, a cumbersome, maintenance-intense delivery system is required.

Sodium hypochlorite however, is effective, easy to handle, easy to store, and the cost makes it the smart choice for chlorine disinfection. The rest of this chapter will be focused on sodium hypochlorite.

Sodium hypochlorite is a yellow liquid. It's usually delivered on trucks, but is also available in drums and carboys. The industry strength is nominally 12.5 percent. It's usually between 12 and 15 percent. Many plants will test the hypo on site in the lab or with portable test kits. I don't accept strengths less than 12 percent. Your organization will almost certainly have an ironclad vendor contract that stipulates minimum required strength.

Hypo is notorious for escaping as a vapor. Operators sometimes call this phenomenon "gassing off." This gassing off characteristic can and often does cause your pumps to air bind, basically placing a vacuum on the system. Sometimes the pumps may need venting. More likely you'll find that sodium hypochlorite feed systems have a venting system built in to counteract the undesirable effects of gassing off. This system usually vents to the atmosphere. Also, the bulk storage tanks are commonly placed inside air-conditioned spaces.

SODIUM HYPOCHLORITE PUMP ROOM

Figure 10-1 depicts a typical hypo pump room. This one is air-conditioned as well, to help with gassing. Behind the pumps along the wall can be seen some of the PVC piping associated with the venting system.

SODIUM HYPOCHLORITE VENT LINE

Figure 10-2 shows a close-up view of the vent line. It's the clear whiteish tubing between the pump and the drawdown tube.

HYPO PUMP AND ASSOCIATED VENT LINE

Figure 10-3 provides a better perspective on the pump and its vent line. The line starts at the bottom on the discharge side of the strainer to the left. It proceeds up the thick pipe with the yellow square painted on it. Follow the line towards the top where it connects to a clear whiteish hose. The hose continues upward where it connects to PVC piping that leads out and is vented to the atmosphere.

Overfeeding chlorine can cause concern for promoting disinfection by-products as well as imparting a noticeable bleach smell in your finished water. Customers will call and complain about this.

Underfeeding chlorine raises the risk of poor to no disinfection. This situation poses a threat to public safety and in extreme cases can put the plant in violation with several federal agencies.

For the record, sodium hypochlorite or "hypo" used in water treatment is basically strong bleach. It's roughly three to five times as strong as store-bought bleach. Be advised, the difference in strength is not to be sneezed at. For example, I once spilled a small amount of hypo on my jeans. I noticed the color starting to fade to white. Okay, I thought, that's what bleach does. I looked again and the spot looked brown. Upon closer inspection, the brown spot was my leg. The hypo had literally eaten a hole in my jeans.

My advice is to observe all the necessary safety precautions when dealing with sodium hypochlorite. Hypo has a relatively high pH and will burn your skin on contact.

Real World 101:

I was changing a broken hose on a hypo pump. I had released the pressure on the system, or so I thought. Apparently the hose was still pressurized. Hindsight being 20:20, I think I forgot to release the rollers. It happens, even to Super Operators.

So I opened the pump casing and released the camlock, a mechanism for unlatching the pump hose from the discharge line. The still-pressurized hose let go and blew hypo all over my face. I was wearing full frontal and side-protected goggles, but within five seconds my face was burning. I went to the emergency eyewash station and activated it from memory and instinct, because your natural inclination is to close your eyes.

Sodium hypochlorite is caustic in nature (it burns on contact). You have to respect it as you should all chemicals. At 12.5 percent strength, sodium hypochlorite is three to five times stronger than the bleach you find in your local grocery store. It will not only turn your pants white; it will literally eat holes in your uniform.

Chapter 11

Ammonia

Ammonia is usually used in the latter stages of the water treatment process, as part of an increasingly popular disinfection method known as chloramination. Specifically, ammonia is combined with chlorine to facilitate the formation of chloramines for disinfection. We'll discuss chloramines in detail in the next chapter.

Ammonia goes by the trade name ammonium hydroxide and its chemical formula is $NH_{3.}$ Water plants normally use a nominal strength of 19 percent. Ammonia has a very distinct and easily identifiable odor. Nitrogen is the main contributor to its pungent smell.

For plants that use ammonia, it's most likely the deadliest chemical in the inventory. It can overwhelm an Operator to incapacitation if the concentration is high enough. The respiratory system is sensitive to and easily irritated by ammonia fumes. Skin contact can result in burns and blistering. The eyes may become irritated and red with prolonged exposure.[1] Fortunately, its pungent odor is offensive to humans. As the atmospheric concentration increases you will naturally want to leave the area post haste (with great speed).

Ammonia storage areas and pump rooms are usually equipped with sensors that measure the percentage of ammonia in the atmosphere. Audible alarms and flashing lights are also incorporated near the sensor and will alert the Operator via the Supervisory Control and Data Acquisition system (SCADA). The SCADA will be addressed later in chapter 20.

Due to ammonia's inherent dangers, additional safety precautions are often employed. Inline air or self-contained breathing systems may be provided as a supplemental safety measure to be used if necessary when handling ammonia. Dedicated ventilation and exhaust systems as well as air scrubbers are common elements that comprise most ammonia delivery and receipt systems. All this specialized equipment points to the serious nature of aqueous ammonia.

Overfeeding can generate customer complaint calls because their water smells of ammonia. Too much ammonia also increases the potential for nitrification in the distribution system, which will significantly reduce your disinfectant residual. Nitrification is the biological oxidation of ammonia.[2] Excess ammonia also disrupts the chlorine-to-ammonia ratio required to produce mono-chloramines.

Underfeeding ammonia can cause the creation of unwanted di- and tri-chloramines. Again, this is due to an improper feed ratio between the chlorine and ammonia. Di-chloramines can be formed as the ammonia is oxidized by the greater amount of free chlorine. The formation of di-chloramines may cause taste and odor problems in the finished water. As underfeeding progresses, tri-chloramines will form. Just like di-chloramines, tri-chloramines are also associated with taste and odor problems. Continued underfeeding will bring you to "breakpoint chlorination."

Breakpoint chlorination occurs when you add chlorine until all the demand has been met and a free chlorine residual begins to form.[2] Chlorine demand comes from anything in the water that wants to consume or bind with chlorine, such as organics and chemicals.

Once the appetite for chlorine is satisfied, then free chlorine forms. The residual chlorine, not having any ammonia to bind with, is now free chlorine. This free chlorine will begin to saturate the distribution system.

Introducing free chlorine into a system using chloramine disinfection can cause discoloration and impart a chlorine smell into the customers' water, resulting in water quality complaint calls.

REAL WORLD 101:

I arrived at work to relieve the nightshift. As I began reviewing the Operators' bench sheet it became readily apparent that towards the start of the nightshift the ammonia had stopped feeding—a fact to which they were obviously oblivious. Table 11-1 shows the relevant data that allowed me to arrive at that conclusion.

Table 11-1 Operator Bench Sheet (excerpt)

TIME	Free Chlorine mg/l	Total Chlorine mg/l	Mono-Chloramine mg/l	Free Ammonia mg/l
2100	3.90	3.87	3.85	0.041
2200	3.76	3.80	3.72	0.038
2300	3.85	3.82	3.80	0.017
2400	3.95	3.74	3.65	0.003
0100	3.82	3.51	3.26	0.000
0200	3.80	2.99	2.68	0.000
0300	3.77	2.14	1.97	0.000
0400	3.92	1.75	1.25	0.000
0500	4.01	1.21	0.75	0.000
0600	4.12	1.98	0.43	0.000
0700	4.18	2.15	0.19	0.000
0800	4.25	3.01	0.002	0.000
0900	4.22	3.37	0.002	0.000

Given the information in Table 11-1 you should be able to interpret exactly what has happened. The free chlorine residual is read just before the ammonia addition. This residual is only one of four inputs that tell the ammonia pump how much to feed. The total chlorine residual represents chlorine that has combined with other chemicals such as ammonia and any unreacted free chlorine. In an efficiently run chloramine system, free chlorine will be zero for all practical purposes. The mono-chloramine residual is calculated by the analyzer and should be very close to the total chlorine residual. The free ammonia should be minimal (between 0.100 and 0.004 is ideal). A small ammonia residual will ensure your ammonia is reacting (binding) completely with the available free chlorine.

As we review Table 11-1, all seems well until 2400. By 0100 you should realize that something is going on. Between 2100 and 0100 your free ammonia residual drops off to zero. At 0100 you can see that your total chlorine residual is dropping and the mono-chloramine residual has dropped off significantly. After 0100 the free ammonia is nonexistent and remains at zero. This is a big clue that you're not feeding ammonia.

At 0200 the total chlorine and mono-chloramine have continued their descent. This is a sign that the free chlorine is taking over, driving all the total chlorine out as it approaches breakpoint chlorination.

Notice the free chlorine residual is rock solid and never wavers. This tells you that you do not have a chlorine problem. Also, note that the free chlorine begins to rise from 0400 through 0800. This is because the Operator sees the total chlorine is in freefall and thinks that additional chlorine will help. The additional chlorine makes the problem worse—a rookie mistake that happened because neither Operator had sufficient knowledge about the plant, especially the chloramination process.

Observe at 0600, the total chlorine is on the rise. The meter is reading free chlorine that has surpassed breakpoint. Mono-chloramine and free ammonia are zero because the ammonia pump has stopped pumping. You can't make mono-chloramine or have an ammonia residual without ammonia.

I performed a drawdown on the ammonia pump and it confirmed the pump was not pumping. I switched to the standby pump. Maintenance was sent to investigate further. They found that the pump had a bad roller head and changed the head in five minutes.

Inadvertently converting a system using chloramines to free chlorine is bad. We'll discuss this in greater detail in the next chapter on chloramines.

References:

1. Aqua Ammonia (5–19.9%), Safety Data Sheet, Air Gas, USA, Radnor, PA, 4/24/2015.

2. *Water Treatment Plant Operation: A Field Study Program*, Vol. 1, 6th edition, Office of Water Programs, California State University, Sacramento, Kenneth D. Kerri, Project Director, pp. 280, 282, 284.

Chapter 12

Chloramination

The process of chloramination is one of the most misunderstood in the industry and can be confounding to Operators. This chapter will bring some clarity to the process. We'll build upon this foundation later in chapter 15 on mono-chloramine analyzers.

Chloramination is a process in water treatment that combines chlorine and ammonia. The process has been around since 1929.[1] The two chemicals are combined to form the chemical disinfectant compound chloramine. Chloramines are a general term that encompasses three (3) variations of the chlorine-ammonia combination: mono-chloramine (NH_2Cl), di-chloramine ($NHCl_2$), and tri-chloramine (NCl_3). When using chloramines for disinfection we want to produce the mono-chloramine species. This is controlled by keeping the chlorine-to-ammonia ratio between 3:1 and 5:1.[2] Outside of these ratios we run the risk of producing di- and tri-chloramines. These are to be avoided because the di- and tri- variants have been associated with taste and odor problems.

Chloramine disinfection is preferable to free chlorine because chloramines provide a longer-lasting disinfectant residual. This is great for helping to keep water in the distribution system safe. Mono-chloramines reduce taste and odor problems, as well as disinfection by-products (DBPs). Fewer DBPs in turn help the plant stay within the Maximum Contaminant Levels (MCLs), set forth in the Stage 1 Disinfectants/Disinfection By-products Rule (D/DPBR).

Although chloramines last longer, they are a weaker disinfectant. This means it takes a higher residual concentration of chloramine to have the same effectiveness as free chlorine. For example, it would take a chloramine residual of 2.0 milligrams per liter (mg/l) to have the same disinfecting power of 0.5 mg/l of free chlorine.[2]

The addition of chlorine and ammonia can occur in one of three application points. You can add ammonia first ahead of the chlorine; this is preammoniation. Preammoniation is not often used because it's not as effective in controlling taste and odors associated with diatoms (a type of algae) and some anaerobic bacteria.

Concurrent application of both chemicals produces the lowest trihalomethane (THM) levels. THMs are an unwanted type of disinfection by-product.

Postammoniation is by far the most common application point, wherein ammonia is added at the end of the process to the plant effluent. This allows for longer chlorine contact time, enabling plants to meet the requirements of the Surface Water Treatment Rule (SWTR).[3]

Overfeeding chlorine, i.e., a feed ratio of greater than 5:1, facilitates the formation of di- and tri-chloramine. Underfeeding chlorine or overfeeding ammonia can set up conditions ideal for the nitrification. I have my Operators keep the free ammonia residual between 0.100 and 0.003 mg/l. Free ammonia residuals greater than 0.1 mg/l should be avoided.[4]

References:

1. Disinfection with Chloramine, Centers for Disease Control and Prevention, cdc.gov/healthywater/drinking/public/chloramine-disinfection.html, January 20th, 2015.

2. Science of Chloramination, Maine Water Utilities Association, Portland District, Michael Koza, June 2010, p. 29.

3. "Chloramination," David Foust, *Water Treatment Plant Operation: A Field Study Training Program*, Vol. I, 6th edition, California State University 2008, pp. 281, 282.

4. Science of Chloramination, Maine Water Utilities Association, Portland District, Michael Koza, June 2010, pp. 18, 26, and 28.

SECTION 3

Plant Hydraulics and the Nature of Pumps

Chapter 13

Plant Hydraulics

Plant hydraulics are an often-overlooked area of water treatment plant operations. Operators may not even be aware of it and how it affects plant operations. Knowing the plant's hydraulics makes you a better and more effective Operator.

What are we talking about when we say "plant hydraulics"? Hydraulics are defined as "The physical science and technology of the static and dynamic behavior of fluids."[1] We're talking about how the plant behaves under various flow conditions, how long it takes chemical changes to respond, and what timeframe your data are in. Are the data two hours old or 24 hours or 15 minutes?

Understanding your plant's hydraulics requires knowledge of the physical locations of various pieces of equipment such as chemical feed tanks, chemical pumps, sensors, monitoring equipment, raw and high-service pumps, etc. You must not only know the individual locations, but also their locations in relation to one another, as you will see.

Knowledge of your plant's hydraulics has many benefits such as:

(a) Helping you know when corrections to treatment should take effect.

(b) Aiding you in deciding the best distribution of in-service filtration.

(c) Providing good information when establishing the best procedures for special evolutions, such as changing clarifiers or sedimentation basins.

(d) Avoiding pitfalls like flooding and loss of treatment.

I don't know the specifics of your plant, but I can tell you some things you should be aware of regarding the hydraulics. You should know how the plant behaves under various flow conditions. For example, one plant I worked at used Infilco Degremont Super Pulsator upflow clarifiers. Understanding the importance of knowing the plant's hydraulic signature, I learned that when bringing a clarifier into service I could get the turbidity to drop very rapidly by filling the clarifier at higher rates of flow rather than lower. For example, filling the clarifier at ten MGD instead of five MGD allowed me to put the clarifier into service in half the usual time.

For plants with multiple filter groups, you might find when regulating basin levels that the levels respond differently to flow changes based on which filter group(s) are being adjusted. The difference in response is usually due to the location of the filter group, the filter size, or a combination of both. All this has to do with the plant's hydraulics.

In the case of plants with different raw pumping stations you may observe varying responses to flow changes or chemical feed changes, depending on which raw pump station is online. To further illustrate the point: I was operating the plant and lost power to the raw pump station due to a contractor's negligence. This plant had two different raw pumping stations. The primary station was located on site and the secondary was off site. I switched to the secondary raw pump station, which was approximately two miles away. It took about 30 minutes to stabilize the flow, and permanganate feed changes took 45 minutes to take effect—all because the raw water now had to travel a much greater distance through varying pipe sizes.

The chemical feed pumps located at the pumping station have their own characteristics. As a result, the permanganate dosage had to be increased significantly to achieve the desired residual. If you're unaware of the plant's hydraulics, it can result in incorrect interpretations of the plant's responses. This in turn may yield poor treatment decisions on your part.

Operating two or more raw pumps in tandem can cause the plant to behave differently as well. I was having difficulty in regulating the applied pH when two raw pumps were online. The difficulty arose because the flow was not stable when the pumps ran together in automatic flow mode. The pumps would "hunt" as they tried to maintain the set flow. Hunting means the pumps were varying greatly, speeding up and slowing down in response to over- and undershooting the target flow. Needing to stabilize the flow, I placed the pumps in manual and split the desired flow (22 MGD) between the two pumps. Location and pump capacity all affect how the plant behaves hydraulically.

It's also important to understand in what timeframe the data lie. Many times, Operators are looking at data that are not in real time, which means the data they're seeing are a result of actions that took place much earlier. This is especially true in plants that are spread out over a large area, with areas that have long detention times. Detention time refers to how long water stays in a specific area such as a clarifier, contact chamber, or clear well. Detention time in clear wells can range from a couple of hours to more than a day in some systems. This must be taken into account when interpreting data and making operational decisions.

For an Operator, this type of information is priceless. A lot of this knowledge comes from observation, talking to Operators that have been there longer than you. This includes seeking information from Operators whose licenses may be lower than yours. I believe you can learn from anyone. If I want to know the best way to wash a car, I talk to the guy who washes the cars, not the owner of the carwash.

Real World 101:

A Class 3 Operator under my charge had correctly made an increase in the chlorine dose going to the settled water collection sump. One hour later he told me he was going to make another increase to the dosage. I asked why. He said because the residual had not increased.

At this point I knew he didn't understand the plant's hydraulics. I explained that the distance between the chlorine pump and the application point in the collection sump was about a quarter-mile. We considered the pump capacity, flow, and the distance the increased dose needed to travel. With this data, we calculated that it would take at least two hours to see any residual increase reflected in his initial dosage change. It takes a small chemical feed pump operating in milligrams per liter (mg/l) quite a bit of time to move its chemical a quarter-mile. We waited another hour and began to see the residual rise. In this case, knowledge and patience kept us from overshooting our desired residual parameters.

References:

1. *American Heritage Dictionary*, 5th edition, Houghton Mifflin Harcourt, Boston/New York, 2011, p. 861.

Chapter 14

The Nature of Pumps

Pumps can be an area of wonder, even though Operators deal with them routinely. Pumps are the cause of a fair share of headache that can occur while running the plant.

Water treatment plants are littered with all kinds of pumps. Therefore, well-rounded Operators should have a basic understanding of pumps, to include the types, classification, and application. You should know when a pump is operating properly and when it's malfunctioning. This chapter is a crash course in pumps, loaded with helpful information.

We use pumps to move material from one place to another. In water treatment that material is usually a liquid, but not always. Sometimes you might be transferring sludge or a slurry of some kind, such as slaked lime or powdered activated carbon (PAC).

Pumps can be put into two basic categories: Positive Displacement and Non-Positive Displacement pumps. Your plant will have both.

Positive Displacement Pumps (PDPs) are used in systems where constant flow is desired and back pressure is unlikely. PDPs should never be deadheaded. Deadheading means closing its discharge path. For example, a sludge pump that feeds into a waste-holding tank. You wouldn't want to the close the fill valve to the waste tank while the pump is running. A positive displacement pump that operates against a deadhead will continue to pump its material until the pressure gets so great that the pipe ruptures, the pump is damaged, or both. Systems with positive displacement pumps usually incorporate pressure-regulating, pressure-relief, or safety valves to relieve excess pressure.

Systems that need a constant flow include: chemical feed, oil, fuel, and sludge transfer systems. These are good examples of where a positive displacement pump would work well.

Positive displacement pumps can be classified as reciprocating or rotary. In either case the principle of positive displacement applies. Some types of PDPs include the rotary lobe, rotary gear, progressive cavity, diaphragm, screw, gear, vane, and peristaltic pumps.

Peristaltic pumps, sometimes referred to as "hose pumps," are becoming increasingly popular due to their low maintenance, low cost, and resistance to air binding. You'll see them routinely used in sodium hypochlorite systems, as well as other chemical feed systems.

Figures 14-1 through 14-6 show some examples of positive displacements pumps.

PERISTALTIC PUMP

Figure 14-1 is installed in a sodium hypochlorite system. Notice the light-colored appendage attached to and extending below the translucent pump casing. The device is a leak detector, found on many peristaltic pumps.

PERISTALTIC PUMP

Figure 14-2 is being used as a sample pump sending prefinished water to the control room sampling sink. This pump is also equipped with a leak detector.

PERISTALTIC PUMP

Figure 14-3 is another peristaltic pump. Notice the heavy-duty construction. These types of ruggedly constructed peristaltic pumps are used to move abrasive and highly corrosive fluids and slurries—carbon, lime slurries, and acids for example.

ROTARY LOBE PUMP

Figure 14-4 is a rotary lobe pump. These pumps are very durable and proficient in transferring viscous fluids such as clarifier sludge.

DIAPHRAGM PUMP

Figure 14-5 illustrates a typical diaphragm-type pump. These pumps are usually comprised of speed and stroke functions. Observe the red knob underneath the "Ammonia 2" label. This knob sets the stroke. To the left of the box labeled ammonia 2 is another small blue box. It contains the rubber diaphragm assembly. Diaphragm pumps are very reliable but are susceptible to air binding and do not do well with volatile fluids such as sodium hypochlorite.

PROGRESSIVE CAVITY PUMP

The progressive cavity pump is a type of screw pump. It's very good at metering very small doses. This pump is being used to feed polymer.

Non-Positive Displacement Pumps (NPDPs) allow for varying flow depending on system demand. These pumps do not have a positive internal seal like their positive displacement counterparts. This lack of a seal allows for slippage inside the pump casing. It's this ability to allow slip that enables you to deadhead a non-positive displacement pump without damaging it. If you closed the discharge valve on an NPDP the pressure would increase and the flow would go to zero even as the pump's impeller continued to turn. As a matter of good engineering practice, I would not deadhead a NPDP for long. Over time the fluid and pump will begin to heat.

NPDPs are found in systems where the demand is variable—potable water systems, for example. By variable demand, I mean that you would probably see a lot of water usage between 6:00 and 9:00 A.M. when people are waking and getting ready for the day. On the other hand, you would see less water usage at 3:00 A.M. when most people are asleep.

Centrifugal and propeller pumps are examples of non-positive displacement pumps. In a water plant, high-service pumps that feed the distribution system are most likely centrifugal pumps.

AXIAL FLOW PROPELLER PUMP

Figure 14-7 is a raw water pump, which must accommodate varying system demand. This pump is an axial flow propeller pump. Axial flow pumps are known for producing high rates of flow as opposed to radial flow pumps, with lower pressure heads.

HORIZONTAL SPLIT-CASE CENTRIFUGAL PUMP

Figure 14-8 illustrates a horizontal split-case centrifugal pump. This one is being used as a high-service pump feeding the distribution system.

VERTICAL TURBINE CENTRIFUGAL PUMP

Figure 14-9 shows a vertical turbine centrifugal pump that is detached from the motor.

Vertical turbine pumps are designed to move water from deep reservoirs and wells.

When operating pumps, we don't want to introduce air into the system. Pumps don't like it. Remember, they're not air compressors. When air gets into the system it can cause the pump to lose suction and worst of all it can cause cavitation.

Cavitation happens when air bubbles form and collapse inside the pump. This is an extremely simple yet accurate explanation. A cavitating pump makes a very distinct sound. The sound is like marbles rolling around inside the pump casing. Cavitation is destructive to the pump and should not be tolerated. When you hear cavitation, find the cause immediately and correct it.

Most pumps represent speed as a percentage and others, like diaphragm pumps, have a speed and a stroke. The speed represents how fast the pump is turning. In pumps that do not use stroke, the pump's output is directly proportional to its speed. The faster it runs, the more it pumps.

Pumps utilizing both speed and stroke operate a little differently. With these pumps the speed still relates to how fast the pump is running. The stroke relates to how much it is capable of pumping. Stroke is also represented as a percentage. Look at the stroke like a bucket. A large bucket can carry more fluid than a smaller bucket. As stroke is increased the pump moves more liquid per cycle. Increasing the stroke setting is like using a larger bucket.

For example, a pump at 50 percent speed and 25 percent stroke pumps 40 gallons per minute. That same pump at the same speed of 50 percent, but with a stroke of 60 percent now moves the fluid at a rate of 65 gallons per minute, even though the speed is the same. The increased pump stroke is like making the bucket larger, thus enabling a greater output without increasing the pump speed. In automatic mode, the program will drive the pump to the appropriate speed and stroke settings.

Real World 101:

Unusual noises like grinding metallic sounds and squealing can be a sign of a failing pump. Other indicators such as low output or nonresponse to dosage changes are signals as well. Pumps that appear to be running at correct speeds and strokes but have low or no output should be investigated. Be especially mindful of oil and water leaks.

Centrifugal pumps usually have a packing gland. The packing gland is filled with packing material and forms a seal around the shaft. Figure 14-10 shows the packing gland housing. Notice the protruding nuts and bolts used to make adjustments when required.

PACKING GLAND

Figures 14-11 and 14-12 show packing glands that are leaking an acceptable amount. The packing gland not only helps seal but also cools the shaft.

Most manufacturers have some allowable leakage, but large leaks (flows) need to be repaired. Normally excessive leaks can be corrected by adjusting the bolts on the packing gland. In more severe cases the packing material will need to be replaced. In either case, these repairs and adjustments should only be performed by qualified personnel.

LEAKING PACKING GLAND "A"

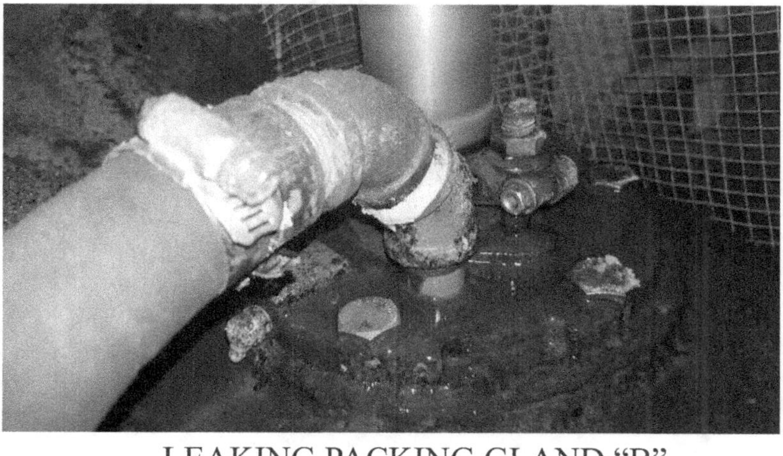

LEAKING PACKING GLAND "B"

SECTION 4

Monitoring Equipment

Section 4 covers some of the most common pieces of monitoring equipment to be found in conventional water treatment plants. Super Operators need to have a firm grasp on what each piece of equipment is for, its basic operation, how to interpret the data, and what operational signs indicate a malfunctioning piece of equipment.

Chapter 15

Mono-Chloramine Analyzers

Mono-chloramine analyzers are used to detect and display several pieces of information vital to maintaining chloramine disinfection. This includes the chlorine-to-ammonia feed ratio, chloramine and free ammonia residuals, as well as the total ammonia being fed.

Chloramines are formed by combining chlorine and ammonia. This process needs to be monitored closely. The online mono-chloramine analyzer is the best piece of equipment for helping the Operator keep the process on track.

The mono-chloramine analyzer is the most complex and misunderstood piece of monitoring equipment in the water treatment plant. Because it's so misunderstood, Operators often misinterpret the data. These misinterpretations can result in improper treatment decisions.

Basic Operation:

Most of these analyzers are microprocessor controlled and use colorimetry to read the sample stream. Think of a microprocessor as a little computer and colorimetry as the measurement of color. A colorimeter is any tool that yields nonsubjective measures of color. Colorimeters also determine a substance's concentration. The concentration is based on the absorbance of the sample. The absorbance is the amount of light that can pass through the sample. The higher the absorbance, the higher the concentration.

Analyzers sample at preset intervals; every two or three minutes would not be uncommon. The sample is fed to the analyzer where reagent is added. The sample and reagent are mixed and sent to the detector (colorimeter) where the color is measured. Many of the substances we're trying to analyze are colorless and can't be seen by the human eye. To overcome that, we add reagents to make them detectable, so they may be compared to the colors of known substances.[1]

The colorimeter consists of a light source, an adjustable aperture (hole), colored filters, some form of cuvette in which to hold the sample, and a detector to measure the color of the sample via the transmitted light. The analyzed sample results can be read on the local display and SCADA.

Mono-chloramine analyzers are equipped with a display and keypad or touchscreen interface for data retrieval, programming, and troubleshooting. The results are displayed in meaningful units such as milligrams per liter (mg/l) or parts per million (ppm).

Note: For our purposes mg/l and ppm are interchangeable.

Mono-chloramine analyzer

Figure 15-1 shows a mono-chloramine analyzer. Notice the local display. Figure 15-2 shows a close-up of the display screen.

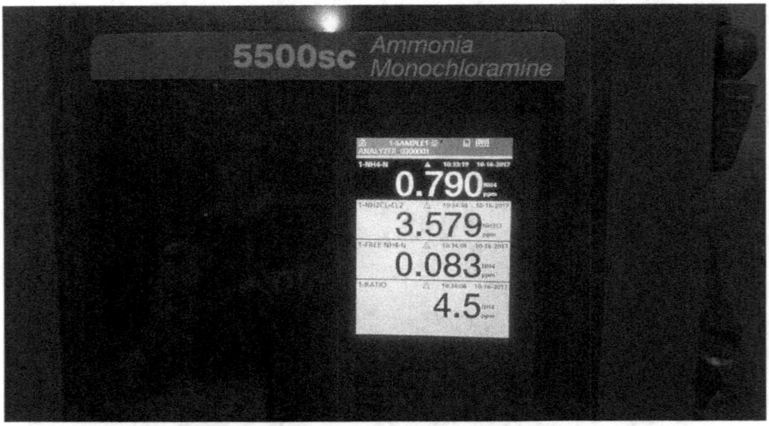

Mono-Chloramine Analyzer Display Screen

A line-by-line breakdown of the display's data follows:

Line 1 is the ammonia feed – 0.790 ppm.

Line 2 is the mono-chloramine residual – 3.579 ppm.

Line 3 is the free ammonia residual – 0.083 ppm.

Line 4 is the ratio of chlorine to ammonia – 4.5 ppm (i.e., 4.5:1).

Note: As a rule of thumb, mono-chloramine residuals should be about the same as the total chlorine residual.

When operating a system using chloramines, you have to know what the key pieces of information mean, and how to interpret the data.

The mono-chloramine residual is the amount of chloramine that is available to be sent to the distribution system. Residuals of three to four mg/l are typical for most plants.

Free ammonia is the amount of ammonia left over after combining with the free chlorine. Most plants want to keep this residual between 0.003 mg/l and 0.100 mg/l. The key is to have some free ammonia left over to ensure a complete reaction, but not so much that it promotes nitrification or can be detected (smelled) in the customers' water.

The free chlorine residual is the amount of chlorine available to combine with the ammonia. The free chlorine residual is normally read independently just forward of the ammonia feed point.

Total chlorine is the sum of the chlorine that has combined with the ammonia and any other chemicals, plus any remaining free chlorine in the water. In a well-run mono-chloramine system, the total chlorine is mostly combined and has little to no free chlorine.

Real World 101:

Let's discuss some common scenarios and talk about what is going on. What are the data telling you?

Case #1 AMMONIA UNDERFEED:

Ammonia underfeed is the most likely scenario to be misinterpreted. Initially you will see a free ammonia residual of zero (0.002 mg/l or less is the same as zero). As time progresses, the free ammonia residual remains at zero. This should be your first clue that something is not right. The mono-chloramine analyzer is screaming that you need to feed more ammonia. Operators sometimes become oblivious to the loss of their free ammonia residual.

No ammonia residual is bad. Left unattended, the mono-chloramine and total chlorine residuals begin to drop as well. Note that your free chlorine residual prior to the ammonia addition is rock solid. Once the mono-chloramine and total chlorine residuals begin to drop, Operators will sometimes increase the chlorine feed. They do this because they see the total chlorine residual dropping, and they panic—thinking they need to increase the chlorine feed to raise the total chlorine residual. The total chlorine and mono-chloramine residuals are dropping because of a free chlorine conversion in progress.

If your mono-chloramine analyzer reads zero free ammonia for two consecutive hours or more, you should investigate for cause. If you doubt your instrument, then by all means run a bench test for mono-chloramine and free ammonia.

Case #2 AMMONIA OVERFEED:

Overfeeding ammonia is just as bad as underfeeding. When overfeeding ammonia, the first sign you will see is an increase in your free ammonia residual. Makes sense, right? Too much ammonia means the residual begins to increase.

The ammonia residual should be regulated to stay within your plant's operating parameters. Most systems accomplish this by adjusting the chlorine-to-ammonia feed ratio on SCADA. For example, changing the feed ratio from 4:1 to 5:1 will cause the ammonia feed to decrease. It should be noted that adjusting the chlorine-to-ammonia feed ratio causes changes in the ammonia feed only.

As the overfeed situation continues, your mono-chloramine and total chlorine residuals begin to drop and your free ammonia residual continues to rise.

The most serious detriment will be releasing excess free ammonia in your distribution system. This excess ammonia tends to promote nitrification.

Nitrification is a biochemical process where nitrifiers (bacteria that react with ammonia) first oxidize the ammonia and form nitrites (NO_2). This nitrite compound continues to react with other bacteria known as nitrobacters, to produce nitrate (NO_3). The nitrates will obliterate the disinfectant residual in the distribution system.

If you notice that the residual in your distribution system rapidly decreases in all or certain areas of the distribution system, you may have nitrification occurring. Especially in areas where these low residuals would be uncommon.

Case #3 CHLORINE OVERFEED:

Chlorine overfeeding, as you can imagine, is not good in a chloraminated system. Too much chlorine will promote the creation of di- and tri-chloramines, which cause taste and odor problems. Discoloration and chlorine-scented water are also undesirable effects of overfeeding chlorine.

As the chlorine overfeed proceeds you will see a decrease in free ammonia and mono-chloramine residual. Left unattended the monochloramine will drop to nothing. The total chlorine residual will also decrease and approach zero milligrams per liter. But it will begin to rise shortly after that. Don't get happy; this is bad. What you're seeing is the free chlorine reaching breakpoint and now you have a free chlorine residual rising. Congratulations, you're releasing free chlorine into a chloraminated system.

Don't laugh; I've seen it happen more than once. Each time the cause was data misinterpretation, except for one instance where the Operator simply ignored the data and in fact never once looked at it the entire shift.

You need to have a thorough understanding of the chloramine process. Closely monitor the chlorine-to-ammonia feed ratio, the free ammonia, total chlorine, and mono-chloramine residuals. By now you should see that effective chloramine disinfection requires thorough understanding and the ability to correctly interpret data.

References:

1. http://www.globalw.com/support/colorimeter.html, Global Water, A Xylem Brand, An Introduction to Colorimetric Analysis & Spectroscopy, 2011.

Chapter 16

Chlorine Analyzers

Chlorine analyzers are important pieces of monitoring equipment and are vital to the proper operation of any conventional water treatment plant. Chlorine analyzers are used to detect residual amounts of chlorine in water. They read free and total chlorine depending upon which reagents are installed.

The analyzers are colorimeters and operate like those discussed in chapter 15 ("Mono-Chloramine Analyzers"). Chlorine analyzers are not as complicated as mono-chloramine analyzers.

A quick review of colorimeters tells us that they measure color and can determine a substance's concentration based on the absorbance of the substance. That relationship is directly proportional: the higher the absorbance the higher the concentration. They take measurements at preset intervals and update the result on the unit's display panel and the plant's SCADA.

These analyzers keep the Operator apprised of the chlorine residuals at various stages of treatment. The chlorine analyzer's importance is elevated because many of the residuals they monitor are federally and state mandated through EPA, health departments, local ordinances, and other such regulatory agencies.

The various regulatory agencies set minimum finished water chlorine residual levels in the plant, distribution system, and the treatment process. These minimum residual levels are monitored, recorded, and reported upon monthly. By law, turbidity records are maintained for a minimum of ten years.

Operators are required to constantly monitor these residual levels. Loss of the chlorine residual in certain instances can place your plant in violation and may even lead to boiled water notices. As such, the analyzers must be well maintained to ensure their reliability.

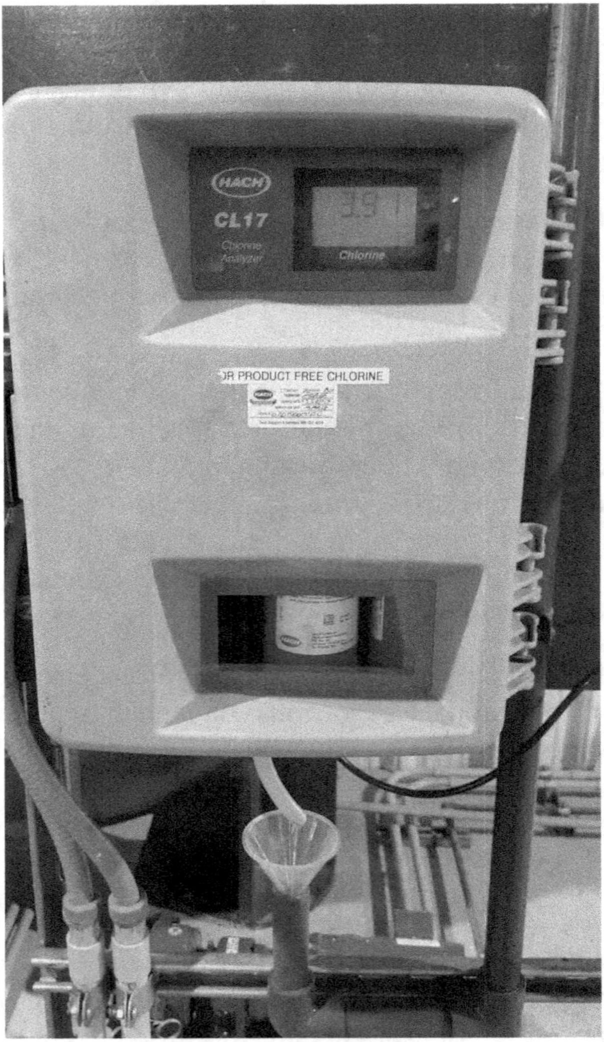

ONLINE CHLORINE ANALYZER

REAL WORLD 101:

Chlorine analyzers can measure very small residuals in milligrams per liter, but they generally become more accurate as the residual rises. The higher the residual, the more accurate the reading.

Most of these analyzers use reagents that must be changed at regular intervals, typically once a month. Indicating and buffering solutions are used as reagents in some units.

If an analyzer begins to read erratically, especially when it's time to change the reagent, check the reagent levels. If you notice discoloration in the reagent bottle, this can also be a sign that the reagent is failing or has been contaminated.

They're also sensitive to interruptions in flow, which can cause temporary spikes or dropouts in your residual reading. You may experience an analyzer that is reading fine and then suddenly drops to almost zero or spikes to its maximum reading and stays there. Often the residual reading will return to normal after several minutes. In these cases, you may have experienced an intermittent flow disruption. If the situation doesn't correct itself, then you may have a more severe problem that requires further troubleshooting.

When thoroughly understood, these analyzers can alert the Super Operator to changes in raw water, failures of other chemical feed systems, and increases in chlorine demand. For example, in a system that feeds permanganate, if the permanganate feed stops you will see a rise in chlorine demand. The rise in demand will cause your chlorine residual to drop. The chlorine residual drops because the organics that would have been oxidized by the permanganate are now free to consume the chlorine. You'll have to increase your chlorine feed significantly to maintain the same residual. Check to see that your oxidizer is still feeding correctly.

Most chlorine analyzers require routine maintenance. Plant personnel can perform the necessary maintenance or it may be covered by a maintenance contract. I'm not a big fan of maintenance contracts. They're usually pricey and not very inclusive. I find that plant personnel do a better job, since they have a true ownership stake.

Chapter 17

Turbidimeters

Turbidimeters are the Crown Prince of monitoring equipment, arguably the Super Operator's first line of defense against treatment plant upsets. Look at them as an early warning system for detecting problems in your treatment.

Before we get started you must have a firm grasp of turbidity. In its simplest form turbidity is the "cloudiness" of the water. The amount of cloudiness or turbidity in your water tells you a lot about the direction your treatment is going. Turbidity is so important that it is monitored constantly through online analyzers and grab samples. SCADA provides for a historical record of it, and plants are legally required to maintain turbidity records for five years.[1]

The technical definition of turbidity is as follows:

> (1) A condition in water caused by the presence of suspended matter, resulting in the scattering and absorption of light. (2) Any suspended solids that impart a visible haze or cloudiness to water and can be removed by treatment.[2]

Turbidimeters measure the cloudiness of water. This measure lets the Operator know the level of suspended solids in the water. For my technical Super Operators, a turbidimeter is "an instrument that measures the amount of light scattered by suspended particles in a water sample, with a standard suspension used as a reference nephelometer."[2]

The basic design of most turbidimeters comprise a light source, a focusing lens, a light trap, and a photoelectric detector. The light source sends beams of light through the sample. The suspended solids reflect this light in all directions. The reflected light is compared against a known standard and the photoelectric detector translates this comparison into a readable measurement called a nephelometric turbidity unit (NTU). The more suspended solids in the water, the higher the NTU measure or turbidity.[3] The turbidimeter's light source is a specific type of light bulb. The light source is the most common single point of failure.

Nephelometric turbidity units are a standard unit of measure for turbidity. The NTU expresses the cloudiness (turbidity) of a sample as measured by a nephelometric turbidimeter.

Figure 17-1 illustrates an online turbidimeter with its associated display unit/keyboard. The display and keyboard are used to calibrate, troubleshoot, and run diagnostic routines.

Online Turbidimeter with Display/Keypad

Figures 17-2 and 17-3 are close-up photos of the turbidimeter and the display/keyboard respectively.

Turbidimeter Assembly

Turbidimeter Display/Keypad

The turbidity measurement of finished water is regulated by the United States Environmental Protection Agency (USEPA). Filter turbidities also have specific turbidity limits set and regulated by the state. In many states, the Departments of Health are actively involved in regulating and inspecting water treatment plants. Most plants set standard operating procedures that are stricter than federal, state, and local regulations to avoid exceeding treatment limits and possible plant violations.

REAL WORLD 101:

Like many pieces of monitoring equipment in water treatment, turbidimeters are sensitive to flow interruptions. Turbidimeters occasionally get slugs of high-turbidity water that pass through them. When this happens, the high turbidity will register but should return to the turbidity range that it was reading previously.

Some indications of failure include displays that are pegged out, reading maximum turbidity— readings that are unrealistic or far different from bench readings. Bench readings are grab samples that are read on an offline or bench turbidimeter (see Figure 17-4). Malfunctioning turbidimeters can also give you very low readings. For example, I would be very suspicious of a filter effluent reading of 0.001 NTU. Filter effluent simply isn't that clear. Lastly, be skeptical of turbidity readings that do not change.

FIG 17-4 Bench Turbidimeter

Most plants routinely clean and calibrate their turbidimeters. Over time the feed lines to the turbidimeter will collect crud and growth in the line. This will cause false high readings as well. These lines should be replaced occasionally, perhaps at six- to twelve-month intervals.

When maintenance procedures are complete, it's important that the unit be assembled properly. For online units, ensure the head assembly is seated properly and no outside light can enter. Outside light will interfere with the unit's operation and give you false (high) readings.

References:

1. Environmental Protection Agency, Title 40, Code of Federal Regulations, 40 CFR 141.33, July 2017.

2. *The Water Dictionary*, 2nd edition, American Water Works Association, Nancy E. McTigue and James M. Symons, Denver, CO, 2010, pp. 629, 697.

3. Visual Encyclopedia of Chemical Engineering, University of Michigan, College of Engineering, Jonathan West, Steve Wesorick, Kelsey Kaplan, http://encyclopedia.che.engin.umich.edu/Pages/ProcessParameters/Turbidimeters/Turbidimeters.html.

Chapter 18

Streaming Current Monitors

Streaming Current Monitors or SCMs are very useful pieces of monitoring equipment. Not all plants use them and many managers and supervisors do not find them useful. I find them valuable and have included this brief discussion.

Simply put, Streaming Current Monitors (SCMs) are used to measure the electrical charge of water. SCMs can be an extremely useful tool when monitoring and adjusting primary coagulant dosage. Determining whether the water has a net negative or a net positive charge will help you fine-tune the coagulant dosage to its most effective level.

A large net negative charge would indicate that you have room to increase your primary coagulant dosage and a net positive charge may indicate an overfeeding of the primary coagulant. Theoretically a net zero charge would mean you have reached the optimum dosage. When running the plant, you may find that your plant runs optimally with a slight negative or positive charge. Each plant has its own unique characteristics.

Streaming Current Monitors can alert you to problems in your treatment. A sudden and significant increase or decrease in the water's charge may point to problems with the primary coagulant. The SCM provides other benefits, such as longer filter runs, less sludge, and reduced coagulant usage to name a few.[1]

Some SCMs use a reciprocating piston assembly and electrodes that read the streaming current. The source of this small electric current is the water being treated. This current is generated as the plant water passes across a surface like the tubing that feeds water to the SCM.[2]

Streaming Current Monitors require routine cleaning. The application will determine cleaning frequency. An SCM reading applied water (raw water with the coagulant and any pretreatment chemicals added) with a turbidity of 50 NTU will need to be cleaned more often than applied water with a turbidity of 3 NTU. I have worked in plants that clean their SCM three times a week. Other plants go an entire month before cleaning the SCM.

References:

1. American Water Works Association, Plant Operations Committee, Optimizing Coagulant with Streaming Current, Chris Griffin, www.vaawwa.org/file/Committees/Plant%20Operations/Online%20Streaming%20Current%20Monitor%20Usage_Theory%20and%20Practice%20.pdf.

2. Chemtrac Inc., Streaming Current Charge Analyzer, www.chemtrac.com/products/duratrac-4-hydroact-series/.

Chapter 19

Flow Meters

Flow meters are integral pieces of monitoring equipment. You will literally find them all over the water treatment plant. No other meters in the plant help to control and provide input to so many systems. A malfunctioning flow meter can literally stop other pieces of vital equipment. Flow meters can cause you to lose treatment and even put the plant in violation.

As the name implies, they are used to monitor the flow of various fluids throughout the treatment process. Some of the flows include raw water, finished water, sludge, and chemical feeds. There are several types of flow meters. I'll discuss some of the types you're likely to encounter in a water plant.

Rotameters:

These simple yet highly reliable flow meters have been around since the mid-1940s. Rotameters are known as gravity-type flow meters. This is because they operate on the theory of the downward force of gravity against the flow of the fluid being measured. For this reason, they must by nature be mounted vertically with the fluid flow in the opposite direction of downward gravity. Rotameters get the name because the early versions used floats that were centered and spun when they were in operation.

Their basic construction consists of a tapered tube and a float. As the flows goes through the tapered tube it drives the float to a certain level within the tube. The assembly has a scale that corresponds to the level of the float. Rotameters are mechanical devices that operate with a high degree of repeatability. They sometimes require occasional cleaning but are not prone to failure. Reliability, low maintenance, and low cost make the rotameter a mainstay among its more sophisticated counterparts.[1]

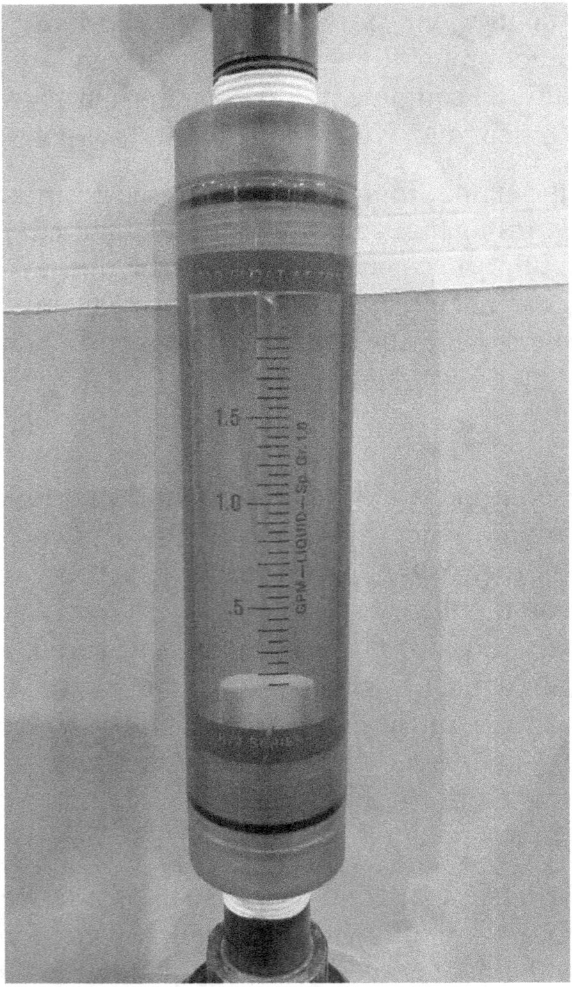

Rotameter

Figure 19-1 is a simple rotameter. Notice the white float assembly at the bottom of the rotameter housing. As the water flows, the float will rise inside the housing. The flow is read by reading the top of the float (see the instructions at the top, "Read Float at Top"). This rotameter is read in gallons per minute (gpm) as indicated on the housing face.

Venturi Tube:

Venturi flow meters are differential pressure flow meters. They're commonly found on filters, especially gravity flow filters. These meters are also highly reliable. The simple, no-moving-parts design makes it a favorite among flow meters.

They take two pressure readings to obtain a differential pressure reading. The first reading is taken from the high side at the wide part of the entrance to the venturi. As the venturi tube constricts, a second (lower) pressure reading is taken at some point downstream in the smaller-diameter portion of the venture, before the tube begins to open back up to its original diameter. The first pressure minus the second gives you the differential pressure.[2] This differential pressure is then converted to a flow rate that is useable for the Operator. Two common units of measure for flows are gallons per minute (gpm) and million gallons per day (MGD). Head loss can also be calculated from the differential pressure of the venturi tube. Head loss is a way of measuring the dirtiness of your filter. The higher the head loss the dirtier the filter.

Venturi Flow Meter

Figure 19-2 is a venturi flow meter used to calculate filter flow and head loss. This venturi is located below the filter in the filter gallery.

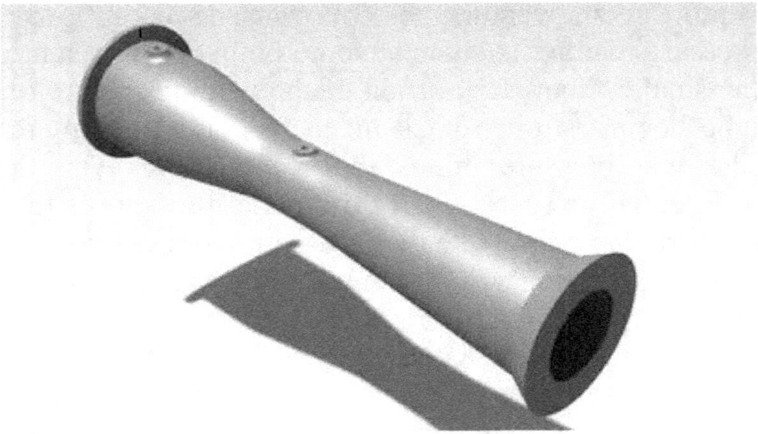

Venturi Tube Basic Shape

Magnetic Flow Meters:

Magnetic flow meters are available in large pipe sizes and are common in systems that carry huge volumes of water, like the plant's main distribution line. These meters have no moving parts and are reliable, but average when it comes to accuracy. You would not place an electromagnetic meter where precise flow measurement is required.

These meters operate on Faraday's Law of Electromagnetic Induction. The flow meter induces a magnetic field into the pipe. A voltage is generated as a conductive fluid, like finished water, passes through the magnetic field. Electrodes in the flow tube sense this voltage and translate to a flow measurement. The faster the liquid moves, the higher the voltage generated, the higher the flow.

The magnetic flow meter has three basic parts: a tube that is placed inline, coils for magnetic field production, and electrodes for reading the voltage induced by the fluid being measured.[3] Nonconductive liquids like deionized and distilled water do not register. Occasionally the electrodes can become coated, affecting their proper operation.

Magnetic Flow Tube Meter

Magnetic Flow Meter, 20-inch pipe size

The meter in Figure 19-5 is from a large transmission line that sends well water to the treatment plant.

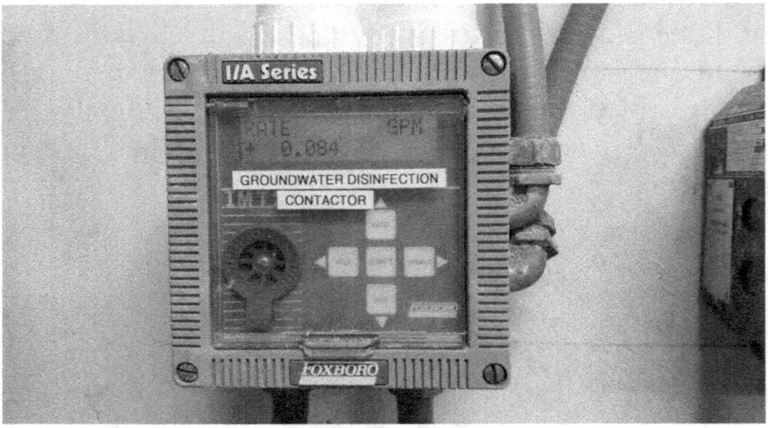

Flow Meter Local Display

Figure 19-6 shows a flow meter's local display. This display also transmits data to the SCADA system and provides inputs to chemical feed systems.

References:

1. The Basics of Rotameters, http://www.sensorsmag.com/components/basics-rotameters, by Sensor Contributor | Oct. 1, 2002 1:00 A.M.

2. Flow Maxx Engineering, Venturi Flow Meters, http://www.flowmaxx.com/venturi.htm.

3. Magnetic Flowmeter Technology, Universal Flow Meters, flowmeters.com/magnetic-technology.

SECTION 5

SCADA

Chapter 20

SCADA Is Your Friend (Get to Know It)

At the heart of all modern water treatment plants, you'll find the Supervisory Control and Data Acquisition system, hereafter known as SCADA. This system is the most underutilized tool in the Operator's toolbox. Its control, diagnostic, data storage, and trending features often go untapped. To maximize the SCADA's full potential, you need to understand how it works and learn its capabilities.

If you only use MSWord to write letters, then you have barely scratched the surface of its true power; so it is with the SCADA. It's imperative that you know the system's capability and how to manipulate it to your best advantage. Unfortunately for most, the SCADA remains a mystery.

An Operator who has a firm grasp of the SCADA is armed with an extremely powerful asset that places an incredible amount of useful data at your fingertips instantly. The SCADA is your friend. Get to know it.

SCADA systems have many benefits, like making plant operations more efficient. Operators can take quick and appropriate actions to keep the plant in compliance. The system has expansive monitoring and trending capabilities that can be used to forecast equipment failures, thus minimizing downtime. Used effectively, SCADA systems save time and money.

Think of the Supervisory Control and Data Acquisition (SCADA) system as the brains of the water plant. Many of the plant's automated functions run through the SCADA and as in any automated system, malfunctions sometimes occur. This is another reason Operators must have in-depth knowledge of how the plant works. The following list helps to further demonstrate functions that SCADA performs to assist the Operator.

SCADA systems provide:

(1) Monitoring and analysis of real-time data.

(2) Control of multiple treatment plant processes remotely, in automatic and manual modes.

(3) Notification of alarm conditions and equipment malfunctions.

(4) Trending for most treatment parameters.

(5) Long-term storage of large amounts of data.

The SCADA is a system of computer hardware and software that allows for the control, monitoring, data collection, and logging functions necessary for the operation of a water treatment plant.

The basic components of the SCADA are an operating system, usually but not always Windows based. Its hard drives have ample storage space and processors fast enough to make the system efficient. Cloud technology is being used in some facilities to handle the ever-increasing data storage requirements. Programable Logic Controllers (PLCs) are scattered throughout the system. The PLCs are microcomputers that talk to a host of equipment in the plant, such as pumps, sensors, monitors, motors, valves, and human/ machine interfaces (HMI). The SCADA network is usually connected through a stand-alone local area network (LAN). A LAN is used to keep the SCADA safe from outside interference, intrusion, or hacking. Most municipalities keep the SCADA intentionally separate from the internet and citywide computer networks.

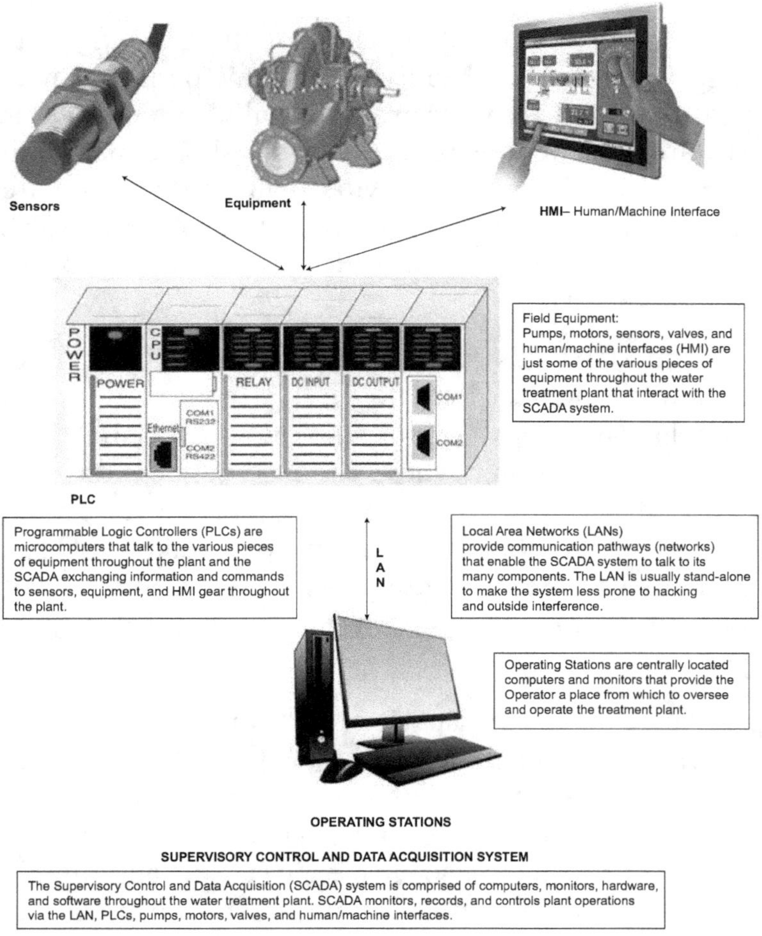

Basic SCADA Diagram

Trending is one of the greatest tools afforded by most SCADA systems. Trending is the ability to graphically and or numerically represent data over specified time periods. Single or multiple data parameters can be trended given the proper commands. This trended data can be an extremely valuable tool for running the treatment plant, predicting seasonal events, equipment failure, etc.

Another important function of the SCADA is the logging of historic data over long time periods, several years in some cases. For example, online filter turbidities are archived for years. Stored data should be readily retrievable when required. Most plants print the required information in report form and store as necessary. The historian function is usually viewed as a backup to these hard-copy reports.

The SCADA system is so powerful that by simply reviewing the data I can logically ascertain everything that has happened during the previous shift. I can pinpoint in real time where various treatment decisions were made, when equipment failures occurred, and when specific corrective actions were taken. SCADA systems can seem daunting, but it is well worth your time to get to know your system in as much detail as possible.

The SCADA is usually broken into neat functional pages that can be pulled up by the Operator for review. Areas such as a general plant overview, chemical feeds, raw flow, and distribution are arranged logically for ease of use. These pages also enable the starting and stopping equipment, or initiating certain automated procedures such as filter backwashes. Each plant will have its own design and screen names. Figures 20-2 through 20-6 show what typical SCADA screens may look like.

The Process Overview screen (Figure 20-2) gives the Operator a "big picture" view of the plant on one screen.

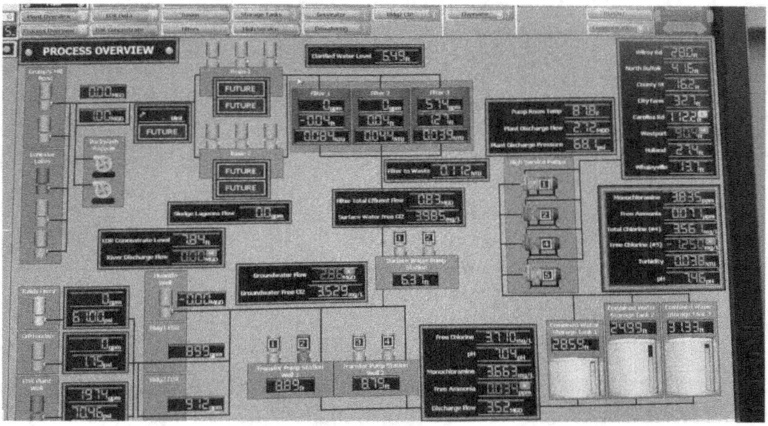

PROCESS OVERVIEW

Figure 20-3 shows detailed information of the high-service system. It includes pertinent data, alarm notification, and control functions.

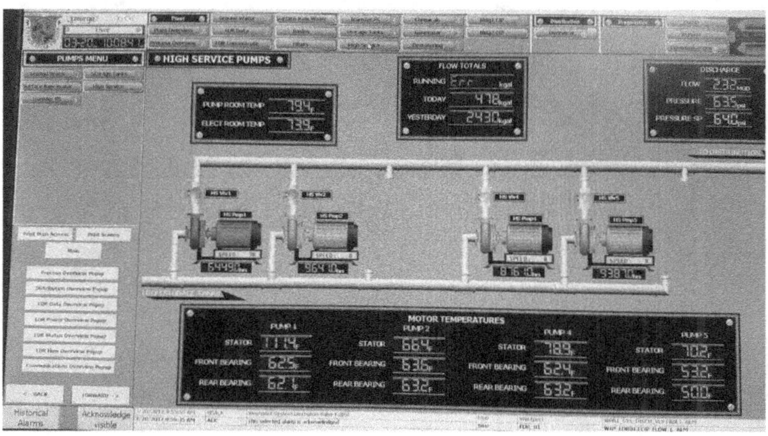

HIGH-SERVICE PUMP PAGE

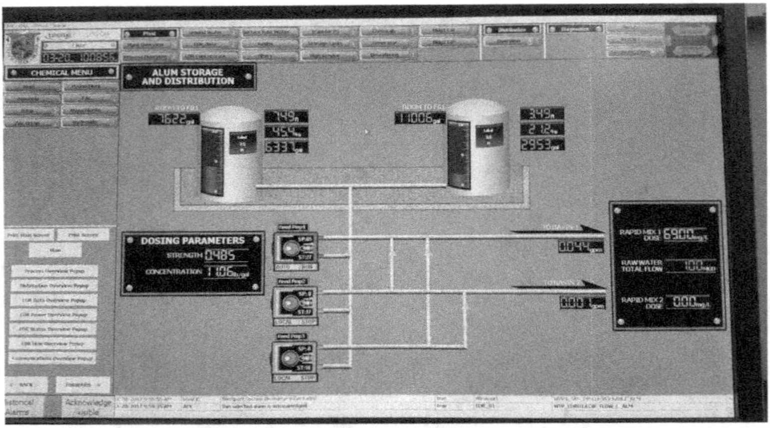

ALUM BULK STORAGE PAGE

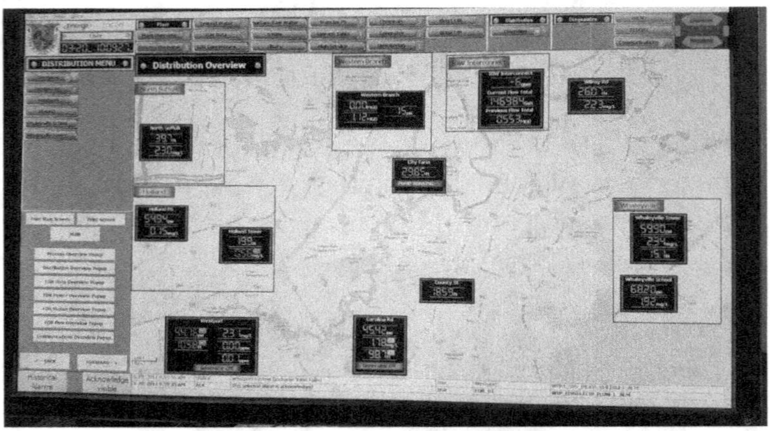

DISTRIBUTION OVERVIEW

Figure 20-6 shows a tank within the overall Distribution System (see Figure 20-5). Figure 20-6 is a sub-page of the "Distribution Overview" page.

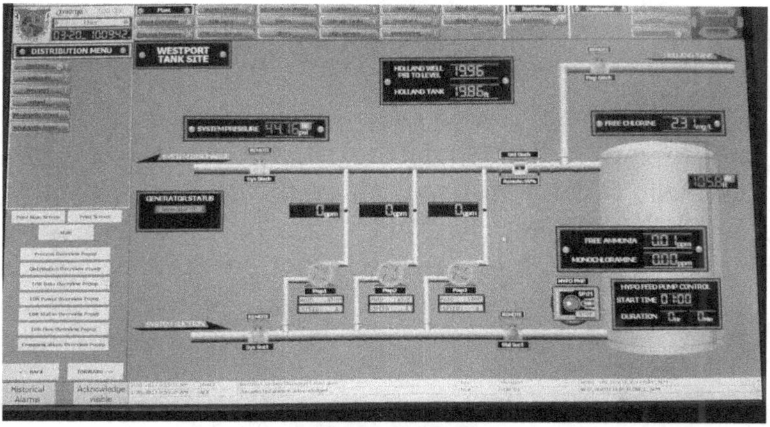

WESTPORT TANK SITE

SCADA pages tell a story and give you an overview of your plant in real time. The Super Operator must learn to understand the relationships between these various pages.

For example, a page that displays various chlorine residuals may show that a certain filter group has an unusually low residual. This may prompt you to pull up the page that shows your sodium hypochlorite feed pumps and check for the proper operation of the pump that feeds that filter group. It may show you an obvious malfunction or something as simple as a low dosage.

When reviewing the various SCADA pages, you must be aware of the time context of the data. If your caustic tank is a quarter mile from your point of application and it takes about two hours for any dosage changes to take effect, you must take this into consideration when interpreting data.

Another vital aspect of the SCADA/Operator relationship that is often overlooked, is that you should know where every sensor, flow meter, CL-17 (chlorine analyzer), transducer, etc. is located and the basics of how it operates. The following situation applies:

When the sonic level reading on a basin suddenly goes to zero, SCADA alarm circuitry alerts you to the low basin level. Knowing where the sensor is located, you walk to it and see that it has been covered with leaves. You also know that obstructions in the sensor's sonic path will give false readings. By removing the leaves, the problem clears and the true reading returns.

Real World 101:

Sometimes trending and historical functions stop or go offline. Left unattended, the loss of these functions can put your plant in violation. It's not always apparent that you have lost these functions. For this reason, it's highly advisable that you learn to recognize when your system's trending and historian functions are not working.

This should be reported to a supervisor and the situation corrected as soon as possible. When these failures occur, most plants have premade log sheets so that readings can be written manually until the system is restored.

SECTION 6

Good Operating Practices

The importance of "Good Operating Practices" may seem to go without saying, but you would be sadly mistaken. Trust me, the chapters included in this section are sorely needed. I have seen many incidents and accidents that occurred due to poor or nonexistent operating practices.

What are "Good Operating Practices"? In a nutshell, good operating practices are simply ways of doing things that are beneficial to plant operations. One example of a good operating practice would be verifying that a chemical tank has enough room to accept the chemical load prior to offloading.

Chapters 21 through 32 discuss various areas of good operating practice. Good operating practices are an attribute that all Super Operators will have in common.

Chapter 21

Reporting to Work in a Timely Manner

As mentioned in the Section overview this chapter may seem self-evident, needing no discussion. As a veteran Operator, I can assure you that you are sadly mistaken. This discussion is vital because an Operator's attendance and ability to show up are critically important.

Water plants, especially large ones, operate around the clock. This means the Operators require a "contact relief." The off-going Operator cannot leave until another Operator arrives on site.

An Operator's job is very different from the eight-to-four, Monday-through-Friday employee. If someone in customer service calls out sick, okay. This means the Customer Service Manager is short one person and the other co-workers pick up the slack. If an Operator calls out sick on short notice, the on-shift Operator can't go home. Another Operator must be called to come in and relieve him or her. The off-going Operator can't get to his sick child or make it to the play on time. You see where I'm going here?

Reporting to work in a timely manner doesn't mean arriving on site with 30 seconds to spare. It means much more than that. It means getting there in enough time to receive a solid face-to-face turnover with the Operator you are relieving.

I recommend arriving at least 10 to 15 minutes prior to the start of your shift. This may sound crazy, but it's what professionals do. Like it or not, Water Treatment

Plant Operators are professionals, licensed by the states in which they operate. You've been granted the sacred trust of the community's health and wellbeing. The serious nature of your job demands that you get to work in a timely manner.

There are a multitude of reasons for being on time. Foremost is because it's the absolute right thing to do. Arriving early gives you time to receive a proper turnover, which is a vital part of assuming the shift.

You should have a firm grasp of the rules and regulations that govern Water Treatment Plant Operators. For example, the part of the Virginia Administrative Code that addresses coming to work in a timely fashion falls under the "Prohibited Acts" section. It addresses the exercise of reasonable care and judgment.[1]

Your organization most likely has written policies and rules that cover timeliness and reporting to work. It's important for Water Treatment Plant Operators to have consistent and reliable attendance records.

Real World 101:

1. Most plants, whether Municipal or Private, generally do not have issues with attendance and reporting to work. With those organizations, if you have a problem with timeliness they just tell you don't worry about coming back the next day or any day thereafter. Poor attendance and lateness simply cannot be tolerated. It upsets the morale among all plant personnel. Discord will start to grow as one person wonders why Operator Smith gets to come in whenever they like. And the "If they can do it, so can I" mentality creeps into the organization.

Get into the habit of leaving for work in plenty of time without having to rush. Rushing leads to errors and mistakes. Arriving 10 to 15 minutes early saves you a lot of headache, frustration, and time. Yes, you save time because you don't have to guess or try and figure things out when you receive a proper turnover. It also puts the off-going Operator in a state of ease, knowing his or her relief is on site in plenty of time to take over and ready to assume the shift. That Operator knows they will be able to meet their next obligation on time. They know you're a responsible, stand-up Operator who understands that their time is just as valuable as your own.

2. I've had Operators literally relieve me late and walk in with a bag of burgers from a fast food restaurant. That tells me they're unprofessional and have little to no regard for me as a person. In situations where an Operator would routinely relieve me late, I have always talked to the Operator candidly, face to face and alone. Without fail they stopped relieving me late.

3. Shifts that have Operators that relieve in a timely manner are more cohesive and have an effective transfer of information between shifts. These shifts are also less prone to operating errors.

4. The consequences of being late or arriving at the last second to assume shift are often far reaching and can carry significant ramifications. Being late will often impact the life of your fellow Operator in ways you cannot imagine. I know an Operator who missed the birth of his daughter because he was relieved over an hour late.

5. In another instance, an oncoming Operator who arrived late put the plant in violation because they received a poor, incomplete, and rushed turnover. The off-going Operator was naturally anxious to leave so he could check on his wife who was in the hospital.

Super Operators report to work in a timely manner because it's a matter of pride and a display of their professionalism.

References:

1. law.lis.virginia.gov/admincode/title18/agency160/chapter30/section320/, Prohibited Acts, Virginia Administrative Code, 18VAC160-30-320, item 6, 1 April 2017.

Chapter 22

Assuming the Shift

The ability to, and habit of, properly assuming the shift is of great benefit and is another characteristic common to all Super Operators. In fact, proper shift assumption is vital. An Operator must be rested and ready to operate when they arrive to take over the plant. The main reason is because operating a water plant requires diligence, razor-sharp focus, and foresight. When operating the plant, you have several systems and a multitude of parameters to monitor. The job requires you to make important decisions—decisions that can have major health and public safety ramifications. You're reviewing lots of numbers and are required to put all these data into coherent context. It may seem easy and second nature, but that's usually when things are running perfect. It's at crunch time when things are going south that you must be ready, and that's when properly assuming the shift pays off greatly.

Being ready to assume the shift and all its responsibilities begins at home long before you leave the house. Make sure you get plenty of rest before your shift starts. Make sure to keep yourself hydrated. Get up in plenty of time to allow for showers, getting dressed, making meals, and taking care of personal matters. Eating nutritious meals at regular intervals is also important. Don't assume the shift hungry. Make sure that you allot yourself plenty of time to arrive at the plant at least 10 to 15 minutes early. Once you get to the plant you should be ready to start the process of assuming the shift. This is not the time to eat breakfast, lunch, or dinner.

Most plants run on 12-hour rotating shifts. Even though most of an Operator's job isn't overly physical, the fact remains that 12 hours is a long time. Especially on nightshift. It's unnatural for the body to be awake in the wee hours of the morning and it works against your body's inner clock. If you are not ready, numbers get blurred over time, you get fatigued quickly, and may even fall asleep. This is when mistakes happen and poor decisions are made.

Following these few pointers will help you arrive in a timely manner ready to properly assume the shift. When you arrive fresh and on point your shift will go by smoothly and it won't seem like every minute is an hour. Your eyelids won't feel like lead weights, ready to close at any moment.

The well-rested Operator runs the plant with a high degree of proficiency. Decisions are spot on, well thought out, and logical. The chances of you putting the plant in violation will be greatly reduced.

Chapter 23

Turnover

You have arrived at the plant, ready to assume the shift. The next step is to receive a detailed turnover from the Operator you are relieving. One of the many advantages to be gained by a professionally executed turnover is that the pitfalls of "not knowing" can be avoided.

Without a solid turnover, both Operators are taking several unnecessary risks. Poor turnover exposes the oncoming Operator to plant upsets and possible violations. The oncoming Operator will be operating in the blind, because they don't have all the information needed to safely and effectively run the plant. The off-going Operator will be in violation of most state statutes requiring the exercise of reasonable care and judgment.

I point to the Virginia Administrative Code (VAC) because I'm most familiar with it. The VAC, Title 18, Agency 160 covers all aspects of the Board for Waterworks and Wastewater Works Operators and Onsite Sewage System Professionals. Specifically, the section covering prohibited acts includes the clause ". . . not demonstrating reasonable care, judgment, or application of the required knowledge, skill, and ability in the performance of the licensee's duties."[1] This covers a wide swath of actions and inactions. Operators who fail to give or receive a proper turnover can be considered to have violated the prohibited acts section of the Virginia Code.

When you're receiving the turnover, some of the main points of interest are: What major pieces of equipment are online, including raw pumps and chemical feed pumps? Are any nonstandard configurations in use, such as having a carbon pump cross-connected by jumper hoses? Has the Plant Manager left any special instructions or evolutions to be executed? What equipment if any is out of service or commission? Who is/are the on-call maintenance person(s)? What work orders have been written? Figure 23-1 is a useful turnover checklist.

Real World 101:

The oncoming Operator gets to the plant two minutes late. Unfortunately, the off-going Operator is mad because she must get to a doctor's appointment with a specialist. It has taken months to get this appointment scheduled.

Understandably upset, she leaves the control room without saying a word as the oncoming Operator enters the control room. The oncoming Operator is late, in a rush, and fails to read the shift report. An hour later he loses number one raw pump.

Since he did not get a face-to-face turnover he is unaware that the two standby pumps have been disabled while being repaired by outside contractors. In the event of such a loss the Operations Supervisor has left instructions to bring the rarely used backup raw pumps online. These backup raw pumps must be manually operated locally at their respective control panels.

The oncoming Operator is new and hasn't been trained to manually operate the backup raw pumps. The off-going Operator should have shown him how to do this. It would have taken less than five minutes.

You will find a turnover checklist (Figure 23-1) that provides a good list of turnover items and questions that should be asked before taking over the shift. This checklist is general in nature but will apply to most conventional plants. It's not all inclusive, so feel free to tailor it as you deem necessary—to fit not only your plant but also your specific style as an Operator.

TURNOVER CHECKLIST 23-1

1. What major equipment is online:

a. Raw pumps and plant flow ______________________
b. Chemical feed pumps ______________________
c. Filters ______________________
d. High Service pumps ______________________
e. Clarifier(s) / Basin(s) ______________________
f. Source water (if more than one) ______________________

2. What recent equipment casualties have occurred:

3. What personnel (managers, maintenance technicians, supervisors, etc.) have been notified of casualty (if required):

4. What work orders have been written on the previous shift (these should also be written in the shift report):

5. Are there any special operating instructions:

* Including but not limited to limits on a particular pump speed, feed rates, dosages, plant configurations to be avoided, etc., etc.

6. Special evolutions to be performed:

* Such as placing a clarifier in service, draining a sedimentation basin, maintaining storage tanks or clear wells at specified levels, etc., etc.

7. On call maintenance personnel and phone number: ______________________

8. Special notifications for Supervisors and Managers:

* For instance call Plant Manager if distribution pressure drops below X psi.

TURNOVER CHECKLIST

References:

1. law.lis.virginia.gov/admincode/title18/agency160/chapter30/section320/, Prohibited Acts, 18VAC160-30-320, item 6, April 17th, 2017.

Chapter 24

Log Review

After looking at all the SCADA screens, log review should be one of the first things you do after assuming the shift. Reviewing the logs can point out trends and abnormalities in your treatment. The logs give you a better picture of the plant.

There are several logs common to most plants, such as the Operator's Shift Report (Figure 24-2), Filter Log (Figure 24-3), Operator's Bench Sheet (Figure 24-4), and the Chemical Drawdown Sheet (Figure 24-5) to name a few. Most plants create these forms locally and tailor them to their needs.

The Operator's Shift Report is a record of all significant events that occur while you are on shift. Anyone reading the Shift Report should be able to tell exactly what has occurred during that shift. Some entries are standard, but here is where you can write anything you think is important. I tell my Operators that you cannot write too much in the shift report. Some plants use loose-leaf paper and others use bound ledgers or notebooks. See Figure 24-2.

OPERATOR'S SHIFT REPORT

DATE: ________ OPERATOR'S ON SHIFT ____________________ TIME: ________

MAINTENANCE WORK ORDERS:

1.______________________ 2. ______________________
3. ______________________ 4. ______________________
5. ______________________ 6. ______________________

REVIEWED BY:

OPERATOR'S SHIFT REPORT

The Operator's Bench Sheet is a form used to record various readings from SCADA, grab sample test results and Operator observations. Items such as chemical residuals, turbidities, pH readings, filter data, and tank levels to name a few are written on the Bench Sheet. Often the Bench Sheet will also have a remarks section for explaining anomalous readings like abnormally high turbidity or for writing in special instructions, especially of a temporary nature. See Figure 24-3.

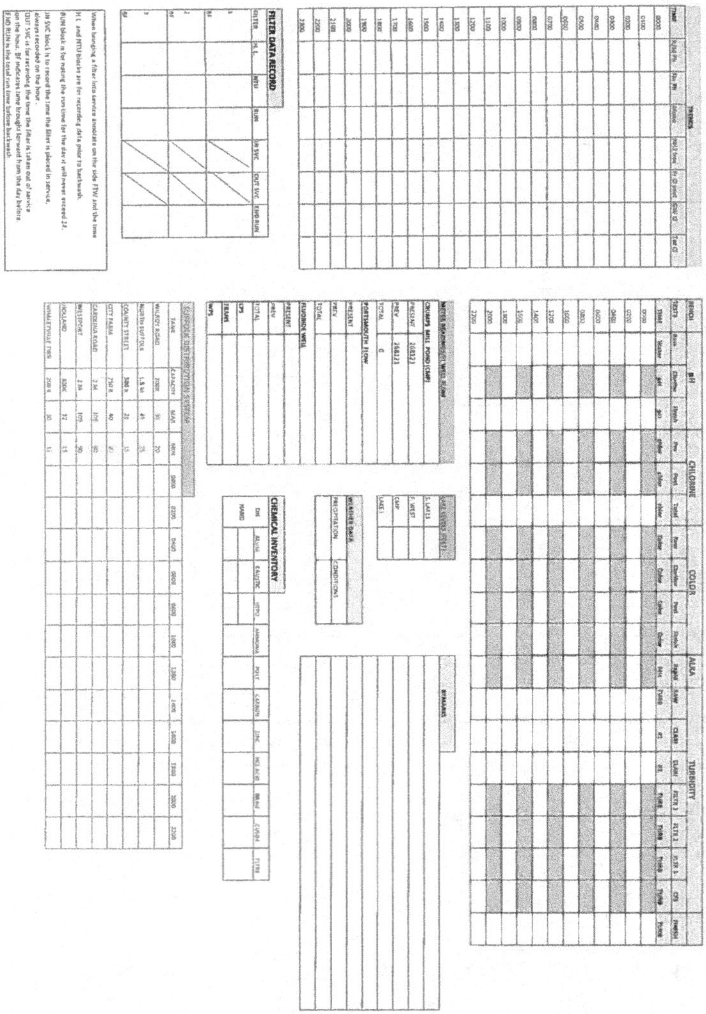

OPERATOR'S BENCH SHEET

The Filter Backwash Log is another log utilized by most plants. It's typically a stand-alone log used to record filter wash data. Much of the filter wash data is state and federally mandated and is included in the monthly state report. The Filter Backwash Log contains information such as total gallons used for the wash, filter run time (how long it was in service) and other data as required. As the Operator in Responsible Charge you should know this log's requirement intimately. See Figure 24-4.

#1

FILTER BACKWASH LOG

DATE	TIME	HDLOSS BF	NTU BF	HDLOSS AFT	NTU AFT	BW GALS	RW MAX NTU	END OF RW NTU	RW TIME (MIN)	FLTR RUN TIME
DATE	TIME	FTW MAX NTU	FTW TIME (MIN)	NTU ONLINE	HEAD LOSS	REMARKS				
DATE	TIME	HDLOSS BF	NTU BF	HDLOSS AFT	NTU AFT	BW GALS	RW MAX NTU	END OF RW NTU	RW TIME (MIN)	FLTR RUN TIME
DATE	TIME	FTW MAX NTU	FTW TIME (MIN)	NTU ONLINE	HEAD LOSS	REMARKS				
DATE	TIME	HDLOSS BF	NTU BF	HDLOSS AFT	NTU AFT	BW GALS	RW MAX NTU	END OF RW NTU	RW TIME (MIN)	FLTR RUN TIME
DATE	TIME	FTW MAX NTU	FTW TIME (MIN)	NTU ONLINE	HEAD LOSS	REMARKS				
DATE	TIME	HDLOSS BF	NTU BF	HDLOSS AFT	NTU AFT	BW GALS	RW MAX NTU	END OF RW NTU	RW TIME (MIN)	FLTR RUN TIME
DATE	TIME	FTW MAX NTU	FTW TIME (MIN)	NTU ONLINE	HEAD LOSS	REMARKS				
DATE	TIME	HDLOSS BF	NTU BF	HDLOSS AFT	NTU AFT	BW GALS	RW MAX NTU	END OF RW NTU	RW TIME (MIN)	FLTR RUN TIME
DATE	TIME	FTW MAX NTU	FTW TIME (MIN)	NTU ONLINE	HEAD LOSS	REMARKS				
DATE	TIME	HDLOSS BF	NTU BF	HDLOSS AFT	NTU AFT	BW GALS	RW MAX NTU	END OF RW NTU	RW TIME (MIN)	FLTR RUN TIME
DATE	TIME	FTW MAX NTU	FTW TIME (MIN)	NTU ONLINE	HEAD LOSS	REMARKS				
DATE	TIME	HDLOSS BF	NTU BF	HDLOSS AFT	NTU AFT	BW GALS	RW MAX NTU	END OF RW NTU	RW TIME (MIN)	FLTR RUN TIME
DATE	TIME	FTW MAX NTU	FTW TIME (MIN)	NTU ONLINE	HEAD LOSS	REMARKS				

NOTES:

GRAY SHADED SECTION IS FOR RECORDING BACKWASH DATA

WHITE SHADED SECTION IS FOR RECORDING DATA WHEN BRINGING FILTER INTO SERVICE

INSERT FIG 24-4 FILTER BACKWASH LOG

Chemical Drawdown Sheets are used in most plants but not all. The drawdown sheet is used to record the actual chemical feed rate of online chemical pumps. The chemical pump drawdown is usually performed once per shift and when a significant dosage or major flow change has occurred. The data from the drawdown can tell you a lot about the pump's performance. For instance, if a pump's dosage is set at 80 mg/l but when you do the drawdown, you find it is only pumping 35 mg/l, then you probably have a problem with this pump. Plants also use the information from pump drawdowns to calculate their approximate chemical usage. The amount of chemicals your plant uses monthly is also reported to state regulatory agencies. Operators sometimes refer to "drawdowns" as "pump calibrations." This is an error. The pump is not being calibrated. You're verifying the pump's actual pumping rate as compared to the set dosage. Figure 24-5 is an example of a chemical drawdown sheet. The dosage and pounds are reportable.

CHEMICAL DRAWDOWN RECORD SHEET

DATE	TIME	RAW	ALUM		HYPO								ZINC		REMARKS
MO/DA/YR		FLOW			#1		#2		#5		#6				
			lbs	mg/l	lbs	mg/l	lbs	mg/l	lbs	mg/l	lbs	mg/l	lbs	mg/l	

CHEMICAL DRAWDOWN SHEET

Of all the logs, the Shift Report and Bench Sheet are among the most important. These should be reviewed thoroughly whenever you take over the shift.

Take your time to read, understand, and interpret what you are reading. The logs are useless unless the data are put into their proper context. Between the Shift Report and the Bench Sheet, you should be able to reconstruct everything that occurred on the previous shift.

Make sure that you read the log from the last time you were on shift. If you were on vacation for two weeks, then you have a lot of reading to do. This point is vital. You only put yourself at a disadvantage if you aren't aware of all that has occurred since you were last running the plant. I was once out for six weeks on sick leave. Upon my return, I read a month and a half of logs and records to ensure I was up to speed. Operating conditions are constantly changing, so you must be acutely aware of the latest information and the history delineating how the plant got to its current conditions.

You must understand that these operational logs are legal records subject to judicial review. If a situation or major violation occurs at your plant, these logs can and will be scrutinized in depth. I have always viewed these logs as an aid and tool for helping me operate in a highly efficient manner. They have saved me more than once, especially in situations where the facts are not being recalled in a totally accurate manner. Another Super Operator told me, "If it ain't logged, it didn't happen!"

Chapter 25

Log Entries and Record Retention

This chapter covers log entries as they pertain primarily to the Shift Report. It also outlines various Operator-related reports and their retention periods.

The upkeep of the Shift Report is a serious matter. The entries should be timely, concise, and accurate. Remember, this is a chronological record of the plant's status, significant occurrences, equipment casualties, plant upsets, etc. Don't forget that the Shift Report and all other logs, files, reports, etc. are legal records.

A wise man once told me, "If isn't logged, it didn't happen." Ultimately the Shift Report tells the reader what happened on the shift. Your entries should have purpose and be necessary. This log is where you can write anything you want. Just understand that what you had for lunch should not be included in the report, unless that lunch has relevance to events pertinent to plant operations.

Real World 101:

1. At a plant that I was once in charge of, I noticed on a Shift Report that every hour, on the hour, was an entry that said, "All plant conditions normal." I immediately stopped the Operators from writing this mindless entry. Shortly after that an Operator asked me, "Well, what *should* I write in the log?"

   I said, "Glad you asked."

Below is a list of items that should be entered in the Shift Report:

a. Date and time of the entire shift, i.e., May 21st, 2018, 0700–1900.
b. The full names of each Operator on duty.
c. All entries should begin with a time entry. Most plants use military time.
d. Starts and stops of major equipment.
e. Start and stop times of major evolutions (i.e., draining a clarifier).
f. Accounts of equipment casualties.
g. Work orders.
h. Significant calls, the subject matter, names, numbers as appropriate.
i. Significant calls you make, such as to the Plant Manager and the details.
j. Special operating instructions for equipment or plant modes.
k. Major flow changes and significant dosage changes.
l. Any actions taken to correct plant upsets or casualties.
m. Significant events such as plant violations, high turbidities, loss of disinfectant residual, etc.

The preceding is a list of recommended items for your report, but it's by no means all-inclusive. Chapter 24, Figure 24-2 is a sample format of a Shift Report. Some plants may use logbooks instead of individual logs sheets.

Operating logs and various other records are required to be kept for certain periods of time. Table 25-1 below lists various records that are required to be maintained and the length of time they are to be held. Table 25-1 refers to Virginia records; however, all states are subject to records management and retention regulations. These requirements stem from several federal laws such as the Safe Drinking Water Act (SDWA) and the Clean Water Act (CWA). Verify the records your state requires to be maintained and for how long.

Table 25-1

RECORD CATEGORY	DESCRIPTION	RETENTION
Waterworks System Monitoring and Regulatory Report	Covers plant monthly operating reports, flow data, chemical use, and analysis records and more.	10 years
Waterworks System Operations Record	Documents plant operations and includes daily, weekly, and monthly logs/reports. Any records covering plant operations.	5 years
Waterworks Individual Filter Monitoring Record	Data, reports, and documents relating to individual filter turbidity.	3 years
Waterworks Operator Continuing Professional Education (CPE) Record	Documents training and education required of all Water Treatment Plant Operators to maintain their certification.	1 year
Waterworks System Inspection Report	Documents the biannual inspection of waterworks systems. Includes raw water, water quality, unit process and other data.	10 years

RECORD CATEGORY	DESCRIPTION	RETENTION
Waterworks Sanitary Survey Record	Required by permit and contains data related to health hazards. Reports, summaries, and correspondence are also included in these records.	10 years
Waterworks Chemical Analysis	Data, documentation, and reports for disinfectant residuals, by-products, precursors, etc.	10 years
Waterworks Bacteriological Record	Includes all bacteriological analysis records like total coliform.	5 years
Waterworks Copper/ Lead Analysis Report	Includes data regarding lead and copper in drinking water.	12 years
Waterworks Consumer Confidence Report	Contains required data for the annual water quality report provided by the waterworks owner to its customers.	3 years
Waterworks System Operating Permit File	Includes the Operating Permit, cover letter from issuer, correspondence with regulator, maps, plans, plant specifications, and treatment processes.	5 years
Waterworks System Operation and Maintenance Manuals	Consists of operations and maintenance manuals for all waterworks facilities and water treatment/ pumping equipment.	Life of the Equipment

The reports and records in Table 25-1[1] are those most relevant to Operators. It only begins to scratch the surface regarding a waterwork's record-keeping requirements. I recommend that you familiarize yourself with the complete list of documents and records that your plant maintains, especially those under the purview of Water Quality. You will be surprised at some of the data required to be maintained.

The Library of Virginia has a website at lva.virginia.gov that holds a wealth of information regarding records and retention. Almost every state has a similar organization that can give insight on the subject.

References:

1. Records Locator Inventory, Library of Virginia, Archival Records Management Services Division, 800 E. Broad St., Richmond, Virginia 23219.

Chapter 26

Plant Overview (Where Am I?)

Once you have assumed the shift, you need to determine where your plant is in real time. This will allow you to figure out the best way to set up the plant. The intent is to put your relief in the best possible position when it's time for you to be relieved. This step also helps you minimize or eliminate the need to make significant changes during the shift.

To figure out where the plant is, you should work backwards. Begin by assessing the status of the distribution system. Note the tank levels. What do the trends and bench sheet data show? Are your tanks dropping, holding, or rising?

Next, look at the high-service system. Your high-service pumps send finished water to the distribution system. Make note of the discharge pressure: Is it normal, high, or low? How many high-service pumps are online? If it's normal to have one high-service pump online but you see there are two online, this may indicate that you're experiencing unusually high demand.

Your clear well(s) is the next area of concern. What are the level(s): full, half-full? Are they rising, falling, or holding?

Then proceed to the filter plant. Which filters are online and what are their flows set at? Do any need to be washed? If so, when will they need to be washed?

Let's check on the chemical feed pumps. You

know the drill: Which ones are online and what are the dosages? Also, take note of which bulk chemical tanks are online and the levels. Do any of them need to be switched immediately?

Now verify what the raw flow is, which raw pump(s) are online. An accurate plant overview is essential for you to set up the plant for the duration of the shift. It will lessen the need for multiple corrections.

As you go through the plant overview process, consider the shift that will be relieving you. Are there special events coming up that warrant the need to take preemptive action? For example, washing filters early because the relieving shift needs to empty a basin and having to backwash filters would unduly constrain those efforts.

Be forward-thinking when you're operating. Taking a critical plant overview will point out glaring shortfalls left by the previous shift and enable you to correct them early.

You may already have a system of jotting notes down or taking an unconscious assessment. This chapter helps you fine-tune that habit and points out why it is necessary.

Figure 26-1 is another checklist provided to help you. I have been operating and running treatment plants for 12 years, and I still use checklists as a tool to run the plant more efficiently.

PLANT OVERVIEW CHECK LIST

DISTRIBUTION:

TANK	LEVEL	STATUS (Rising, Holding, Dropping)/ remarks
________	____	________________
________	____	________________
________	____	________________
________	____	________________

HIGH SERVICE SYSTEM:
ONLINE PUMPS ______________DISCHARGE PRESSURE _______ psi

CLEAR WELL STATUS: FULL ¾ ½ Low Remarks: ____________________

FILTER PLANT:

ONLINE FILTERS	FLOW	HOURS	TURBIDITY	REMARKS / WASH NEEDED
________	____	____	______	____________
________	____	____	______	____________
________	____	____	______	____________
________	____	____	______	____________
________	____	____	______	____________

CHEMICAL FEED PUMPS:

PUMP	DOSAGE	PUMP	DOSAGE	PUMP	DOSAGE	PUMP	DOSAGE
____	______	____	______	____	______	____	______

PUMP	DOSAGE	PUMP	DOSAGE	PUMP	DOSAGE	PUMP	DOSAGE
____	______	____	______	____	______	____	______

BULK TANKS ONLINE:
_______ _______ _______ _______ _______ _______ RMKS: _______

RAW PUMPS ONLINE:

PUMP(S): ___________ FLOW: ________ RMKS: ______________________

NOTES:
____
____
____
____
____
____
____
____

Chapter 27

Operational Assessment

(Getting Where I Need to Be)

Most plants have a philosophy of operating. The simplest and most common is the practice of filling storage and distribution tanks at night and dropping them in the day. This not only provides for tank turnover but leaves the oncoming night shift Operator a place to put the water being treated. These operating doctrines are usually unwritten but are known to the Operators. They dictate how the plant is run in terms of volume of water produced during the shift.

No Operator likes to come on to the night shift and see the distribution tanks are nearly full. He knows that at two o'clock in the morning when everyone is asleep and no one is using water, the distribution tanks will be full and he'll have no place to put the water.

Chapter 26 told you about how to assess where your plant sits in real time. Now that you know where your plant is, this chapter will help you figure out the best plan for turning the plant over to the oncoming shift, getting where you need to be.

Real World 101:

This Real World 101 example demonstrates the process. Let's assume that your plant operates by filling tanks at night and dropping them during the day.

You come in to assume the nightshift. You've gotten an overview of where your plant is and the following is an abbreviated summary of what you know:

The distribution tanks are a little lower than half full and dropping. The high-service system has one pump online at about 65 percent speed or about seven mgd. Your clear well is full and still rising. There are four filters online, all set at maximum flow and holding your clarifiers at optimum levels. All the requisite chemical feed pumps are online and operating automatically. One raw pump is running in automatic at less than 50 percent capacity, roughly 8 million gallons per day (mgd).

Now you must figure out where you need to be. In this case, you need to have all the distribution system and clear wells full when the dayshift comes on in the morning.

Based on the above information you set up your plant. Steps 1 through 5 walk you through the process of getting where you need to be.

Step 1

You must stop your distribution tanks from falling. It's time to fill up, and there are fire protection concerns to take into consideration. Fortunately, your clear well is full and you don't want it to overflow. This is a waste of money and perfectly good drinking water. The one high-service pump you have online is sufficient. You need to raise the speed to increase output and overcome the dropping trend in the distribution tanks. In this case we'll increase the high-service pump output to ten mgd. The goal in raising the high-service pump speed is to start adding more water to the distribution system than you're taking out. Once you achieve this you'll see the tanks begin to rise. Simultaneously you'll start to draw more water from the clear well, causing it to drop. This is good, because your clear well is close to overflowing.

Step 2

Now it's time to address the raw flow. Is it high enough to sustain you throughout the night? A good rule of thumb is that if you are sending more flow out to distribution than you are into the clear well, over time you're going to lose ground. In this case, you are now sending ten mgd to distribution after increasing the high-service pump speed. Your raw flow at eight mgd will not sustain you throughout the night. Considering that later you must reduce your flow to distribution, you raise your raw flow to nine mgd. This will keep your clear well from hemorrhaging throughout the night.

Step 3

Can my current online filters sustain the new raw flow increase from eight to nine mgd? Since your online filters are already at maximum flow, you will need to bring another filter online to regulate clarifier levels and keep them from overflowing. I would bring the fifth filter into service before increasing the raw flow. Good engineering practice would be to balance the filter flow as evenly as possible. However, knowing your plant's hydraulics and individual filter capacities will help you decide how to balance flow.

Step 4

Verify that your chemical pumps that operate automatically have responded to your changes, especially the raw flow change.

Step 5

Monitor the plant to ensure it is proceeding in the direction you need it to go. In this case your distribution tanks should begin rising, and your clear well(s) should be dropping slowly during the early hours of filling. Towards the end of the shift as tanks are getting full, you reduce your high-service pump speed, because you need less water in distribution. This speed reduction will cause your clear well to begin rising. If required, adjust as necessary. At the end of the shift your storage and distribution tanks should be at adequate levels.

As you learn your plant and how it responds, you'll become very good at setting up at the beginning of your shift. I can come into my plant, and once I set it up initially only minor corrections are required. Setting up your plant at the beginning of the shift makes for a smooth transition when the oncoming shift arrives to take over.

Poor plant overviews and assessments can cause you to make rash and dramatic changes. For example, having to make large raw flow changes of more than two mgd, to make up for shortfalls, is usually the result of bad decisions at the beginning. Bad setup in the beginning can lead to other ill-advised choices that have resulted in filters being drained below the bed, the emptying of clear wells, and total plant shutdowns in extreme cases.

See Figure 27-1 for an Operational Assessment Checklist.

FIG 27-1

OPERATIONAL ASSESSMENT CHECKLIST

SHIFT:DAY ____ NIGHT _____

END POINT: FULL TANKS ____ TURNED OVER TANKS ____

CLEAR WELL(S): FULL _____ DROP FOR TURNOVER ______

DISTRIBUTION TANKS: FILL ____ DROP ____ HOLD ____

HIGH SERVICE PUMPS:

INCREASE FLOW ____ REDUCE FLOW ____ ADD PUMP ____

RAW FLOW: INCREASE ____ HOLD ____ REDUCE ____

FILTER PLANT: HOLD FLOW ____ INCREASE FLOW ____ REDUCE FLOW ____

Milestones:

____
____
____
____
____

Remarks:

____
____
____
____
____
____
____

OPERATIONAL ASSESSMENT CHECKLIST.

Chapter 28

Touring the Plant

Touring the plant is a simple but overlooked, neglected, and often not even thought-about practice. A plant tour is what the phrase implies. It means walking through the entire plant. This practice has saved me many times in the past. I have caught leaks in their infancy and averted possible catastrophe.

Touring the plant is an inspection in its simplest terms. You're looking for leaks, housekeeping issues, flooding, abnormal noises, system misalignment, burned-out light bulbs, etc. Basically, anything out of the ordinary. The tour should take you through every part of the plant, especially the dark and wet places that no one likes to go to. Places like the filter pipe gallery and any place in the lower levels. Each building, loading dock, electrical spaces, and high-service pump room should be toured. Make a complete tour at least once per shift.

Real World 101:

Now let's look at some real-world cases where a plant tour has averted disaster.

Case 1:

While touring the basement in the dreaded filter pipe gallery, I saw water as I stepped onto the bottom of the staircase. This immediately set off alarms, because there should not have been water there. As I approached the threshold I saw what the problem was. The sump pumps in the lower level had stopped working. There were no high-water or flood alarms installed, and left unattended the water would have flooded the entire lower level. As this case goes, I caught the situation in time. There was about a foot and a half of water covering approximately 600 square feet. Fortunately, the water had not reached any instrumentation, control panels, or electrical wiring. Maintenance was called, and they cleaned two stuck floating-ball check valves in the submersible pump discharge piping. Sump pump operations were restored and the water was removed.

Case 2:

Walking towards the high-service pump building and about 75 feet before I get there, I can hear a loud, unusual noise. As I open the door to enter I immediately know what's wrong. The online pump sounds like there are marbles rattling inside the pump casing. The pump is cavitating. This is very bad for the pump and will damage it if not corrected. In this case I opted to start a standby pump and shut down the cavitating pump. I notice the pump started leaking water at a flange on the suction side as it was stopping. More than likely, the gasket between the flanges had a blown gasket and was sucking in air—which can lead to cavitation.

Case 3:

Touring the chemical building, I observed a pool of greenish-yellow liquid underneath the discharge hose of an online sodium hypochlorite (hypo) pump. The droplets of hypo were enough to tell me that the hose was leaking. Of course, I started a standby pump, stopped, and isolated the leaking pump.

Now it doesn't take a genius to figure out what the problem was. But that's not the point. The point is that without a deliberate and thorough plant tour, you probably would not have caught the aforementioned cases until they became big problems.

Figure 28-1 can help you to keep track of discrepancies found during the tour. Some issues may require writing maintenance work orders, while others like housekeeping matters simply need to be noted and corrected later on the shift.

FIGURE 28-1

PLANT TOUR CHECKLIST

DATE: ______________ TIME: ____________

SPACE TOURED:	REMARKS:	WORK ORDER: Y / N
____________	________________	____
____________	________________	____
____________	________________	____
____________	________________	____
____________	________________	____
____________	________________	____
____________	________________	____
____________	________________	____
____________	________________	____
____________	________________	____
____________	________________	____
____________	________________	____
____________	________________	____

PLANT TOUR CHECKLIST

Hopefully this chapter will impress upon you the importance of touring your plant at least once per shift. It will save you quite a bit of headache in the long run and at the end of the day, it's the professional thing to do.

It's what Super Operators do.

Chapter 29

Aligning Systems

To align is to arrange things so that they form a line, or are in proper position. I once told a plant manager that I needed to verify the system alignment before starting a pump. He said curiously, "Verify *what?*"

I said "alignment," and then explained that I needed to make sure all the valves were in the proper position and ensure that the pump mode was set up for automatic operation. He looked like he had just had some grand epiphany and said, "Yeah, that's a good idea."

Improper system alignment is the number one cause of self-inflicted equipment damage. All Operators at one time or another have misaligned a valve or left a pump in manual instead of auto, causing the plant to get out of specification. Oh, you never accidently left that alum valve closed? Yes, you have, and if you haven't it's going to happen. System misalignment happens to every Operator; it's just a matter of time.

Ensuring your systems are properly aligned can prevent equipment damage, minimize plant violations, prevent flooding and a host of other undesirable, totally avoidable situations. Double-checking a system's alignment before starting equipment or transitioning the plant into different configurations is imperative. This practice has saved me more times than I can count.

I use and recommend the hand-over-hand method of system alignment. This method is simple and easily learned. Most water plants lend themselves to this method because the pipes, valves, pumps, tanks, and control panels are out in the open and easily accessible.

Most plants have Standard Operating Procedures (SOP) and system diagrams that can be used in conjunction with the hand-over-hand method. The following paragraphs describe a scenario and illustrate how the method is used. It is not intended to replace your plant's SOP or instructions.

Real World 101:

The hand over hand method is described in the following real-world scenario below:

Background:

The #1 sodium hypochlorite pump feeding the online filters has suffered catastrophic failure, due to burned motor windings. The #3 pump is available and can feed multiple application points including the online filters.

Solution/Task:

You must align the #3 sodium hypochlorite (hypo) pump for automatic operation from SCADA, restoring the hypo feed to the online filters.

Note 1:

Since the pump can feed multiple application points, you must ensure it is properly aligned to send sodium hypochlorite from the bulk storage tank to the online filters in automatic mode. Make sure that you are not inadvertently feeding other application points.

Hand-over-Hand Method:

Note 2:

This example is for illustrative and informational purposes only. It is not intended to supersede plant Standard Operating Procedures. Figure 29-1 below has been provided to better illustrate the hand-over-hand method in the following scenario:

FIGURE 29-1

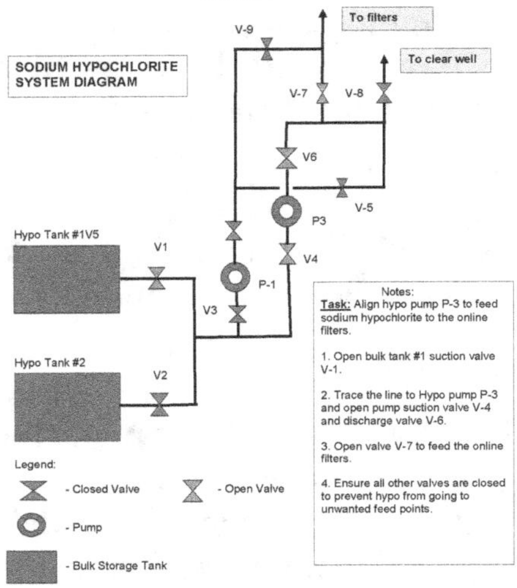

SODIUM HYPOCHLORITE SYSTEM DIAGRAM

Step 1:

Define your starting and ending points. In this case, we start from the online bulk hypo storage tank and end at the feed point for the online filters.

Step 2:

(a) Start at the bulk sodium hypochlorite tank (Hypo Tank #1) and open the tank suction valve (V1). The suction valve is the first valve between the tank and the system to which it supplies.

(b) Continue to follow (tracing hand over hand) the pipe from the bulk tank suction valve (V1) to sodium hypochlorite pump #3, opening any valve that is in a direct path to hypo pump #3.

(c) Once you arrive at hypo pump #3, open the suction valve (V4). This is the first valve on the suction (inlet) side of the pump.

(d) Next, open #3 hypo pump discharge valve (V6). This is the first valve on the discharge (outlet) side of the pump.

(e) Continue to trace the line hand over hand toward the valve(s) that lie in the direct path to the online filters feed point and open it. Valve 7 (V7) feeds the online filters.

Note 3:

You must continually bear in mind that all other feed points from hypo pump #3 should be closed. Simply put, you want to prohibit the hypo from going places you don't want it to go.

(f) At the pump's local control panel, ensure it is aligned for automatic remote operation from SCADA. This usually involves placing a switch into the remote or automatic position.

Congratulations! You have just completed a "Hand over Hand" alignment of the #3 sodium hypochlorite pump for automatic operation from SCADA, feeding the online filters. Whew!

This method is the best for new operators and those not familiar with the system. The hand-over-hand method reduces the possibility of misalignment. New Operators should consider having a senior Operator or someone familiar with the system verify your alignment.

Chapter 30

Receiving Chemical Loads

Receiving chemical loads is a very routine and mundane task. Nevertheless, it is a task that can lead to disaster if Standard Operating Procedures are not followed.

I once asked an Operator who had come into the control room from taking a load of alum, "Do you have enough room in the tank to hold the load?" The answer was "I guess so. I just assumed since somebody ordered it, that there was enough room." I was flabbergasted.

Overflowing bulk chemical tanks is universally bad, but it happens more often than you think. Most chemical tanks have an overflow system. These tanks usually overflow into some type of containment area that is open to the atmosphere. Chemical overflows can lead to neighborhood evacuations, unwanted television interviews, and a host of negative outcomes.

This chapter addresses some good engineering practices that I recommend when receiving bulk chemical loads. These recommendations are not meant to supersede your plant procedures, guidance, or directives.

There are a few things you should always do when receiving chemical loads. First, when a load arrives it should not be a surprise. It should have been passed down or posted, so that you are aware of the delivery.

A sharp Operator will make note that a delivery is coming, verify the tank to be filled, and ensure that it can hold the entire load. If the tank doesn't have enough room, then often the load is split between two tanks, provided the chemical has more than one bulk storage tank. Again, you must calculate if there is enough room to split the load. Also, determine if the tank you are filling is on suction (online). Filling online tanks is not recommended.

When the delivery truck arrives, you should verify that the chemical being received is what you are expecting. This sounds elementary but I'm telling you, Operators have offloaded chemicals into the wrong tank. Many of the chemicals we use have specific strengths—for example, 25 percent caustic soda and 12.5 percent sodium hypochlorite. You must ensure that it's not only the right chemical, but that the chemical meets the specifications required by your plant. Bulk chemical distributors have sometimes delivered 50 percent sodium hydroxide (caustic soda) instead of the standard 25 percent. If this occurs I would contact my Supervisor immediately and not receive the load until I got further guidance.

The driver should have at least three pieces of paperwork:

(a) The Bill of Lading:

Which is not a bill at all. The bill of lading tells you what you are getting. It should have the name of the chemical and the quantity. The quantity may be approximate, but that's alright. Remember, this is not a bill. In fact, its main purpose is to document what you have received. See Figure 30-1.

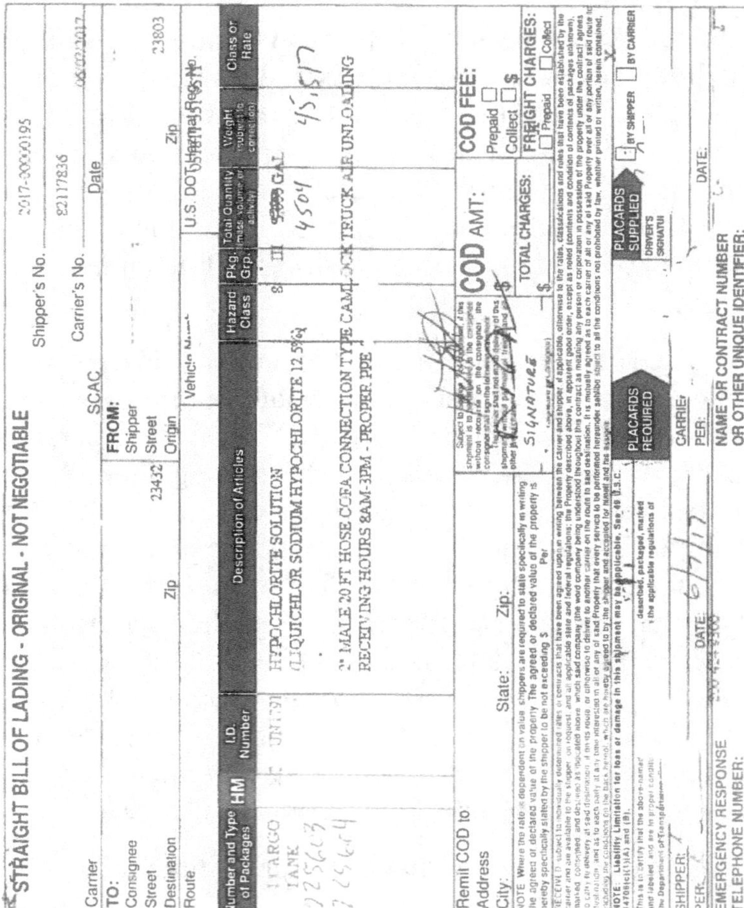

BILL OF LADING

(b) The Weight Ticket:

This is the document that tells you how much chemical is being delivered. Conventional weight tickets will have a total weight, tare weight, and a net weight. The total weight is comprised of the truck and chemical weights combined. The tare weight is the empty weight of the truck. The total weight minus the tare weight gives you the net weight. The net weight is what you will be charged for. Recently, weigh stations are giving drivers weight tickets with the volumetric quantity listed on the ticket. Volumetric weight tickets are perfectly fine and are very accurate. See Figure 30-2, which details an example of a volumetric weight ticket, and Figure 30-3, which shows an example of a conventional weight ticket.

VOLUMETRIC WEIGHT TICKET:
(EXAMPLE ONLY)

ICELANDIA
VOLUMETRIC VERIFIED

TRUCK BAY 9

23 SEP 2018 11:23:47
TOTAL: 4495.0 US GALS

Batch number: 103782

SCALE TICKET

VOLUMETRIC WEIGHT TICKET

Example conventional weight ticket:

WEIGHT TICKET

ACME CHEMICAL INDUSTRIES
101 ANY STREET
PICKA, STATE
00000

Purchaser / Customer : Any City USA
101010 Main Street
Unamethe City, Anystate 00001

Commodity: ____ Ammonium Hydroxide
____ Anhydrous Ammonia
____ Other

Truck/ Tractor: 943az GROSS 60250 lb
Trailer: 50019zbt-01 TARE 29500 lb
NET 30750 lb
05:47 AM 11/30/2012

Driver ___ ON ___ OFF WEIGHED BY _______ LICENSE _______

CONVENTIONAL WEIGHT TICKET

(c) Quality Analysis Sheet:

The Certificate of Analysis (COA), also called a Quality Analysis (QA) sheet, gives you quite a bit of information about the chemical you're receiving. It contains data about the chemical, such as strength, specific gravity, alkalinity, and appearance. See Figure 30-4 for a sample Certificate of Analysis.

Certificate of Analysis

Water Treatment LiquiChlor®

Sodium Hypochlorite Solution

Lot #:	158-17-1036-E	Customer:	
Ship Date:	6/7/2017	Customer PO:	
Trailer #:	61951730	Bill of Lading#:	
Facility:		Univar S/O:	2017-00000195

Analysis	Specification	
Chlorine, available % by weight	Report	12.66
NaOCL, % by weight	12.5% min	13.30
Free Alkalinity as % NaOH	0.25-0.80	0.71
Specific Gravity	Report	1.211
Visual Appearance	Clear to Pale Yellow W/O Suspended matter	PASS

Signature

See label and product bulletin for additional information. Follow EPA approved directions for use. This product conforms to ANSI/NSF 60 Drinking Water Treatment Chemicals - Health Effects

NSF

Maximum Use Level 99 mg/L

Certified to NSF/ANSI 60

For additional information contact: Univar USA, Inc. 1001 Old Bermuda Hundred Road, Chester, VA 23836 804 748-8100

CERTIFICATE OF ANALYSIS

(d) Invoice:

One last piece of paper that's connected with every chemical load is the actual invoice. The invoice is the bill. The driver will not have the invoice. It comes later usually through the mail. (Don't ask me why invoices are still delivered using snail mail.) Ultimately the invoice makes its way to the plant to be processed for payment. The invoice will have the exact quantity that was received, and that quantity should match what is on the weight ticket. It will have the price per unit of measure, for example, \$1.25 per gallon or \$0.50 per pound. It should also name the shipper and receiver, the invoice number, and maybe a purchase order number. See Figure 30-5 displaying a sample invoice.

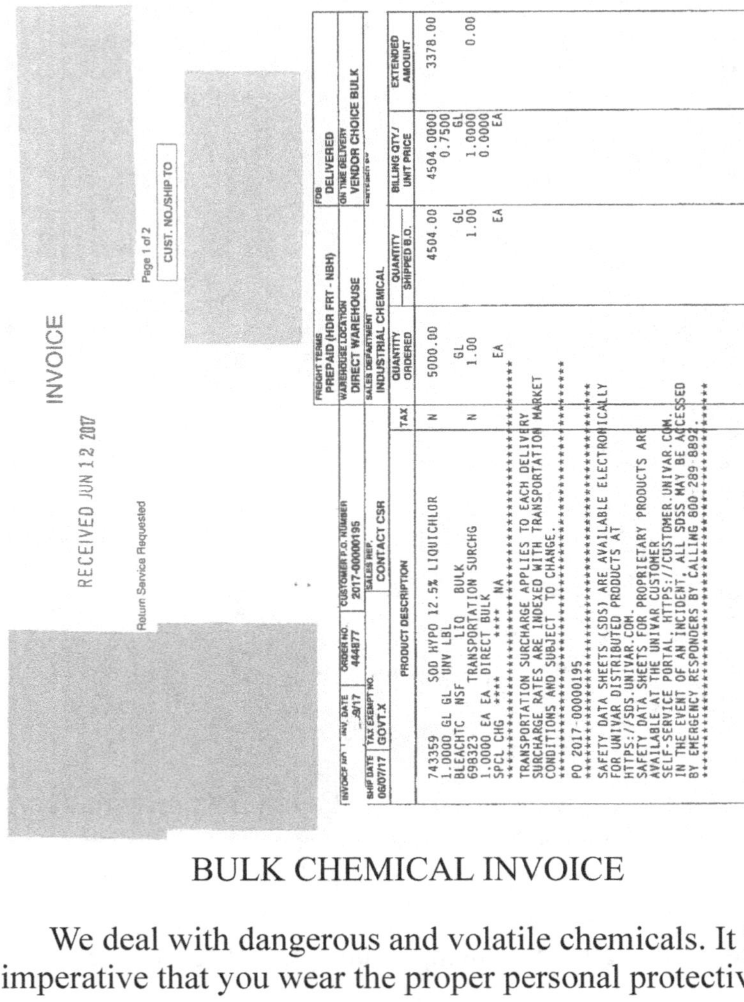

INVOICE

RECEIVED JUN 12 2017

Return Service Requested

Page 1 of 2

CUST. NO./SHIP TO

INVOICE NO.	INV. DATE	ORDER NO.	CUSTOMER P.O. NUMBER	FREIGHT TERMS	FOB
	..9/17	444877	2017-00000195	PREPAID (HDR FRT - NBH)	DELIVERED
				WAREHOUSE LOCATION: DIRECT WAREHOUSE	ON TIME DELIVERY: VENDOR CHOICE BULK

SHIP DATE	TAX EXEMPT NO.	SALES REP.	SALES DEPARTMENT
06/07/17	GOVT.X	CONTACT CSR	INDUSTRIAL CHEMICAL

PRODUCT DESCRIPTION	TAX	QUANTITY ORDERED	QUANTITY SHIPPED B.O.	BILLING QTY./ UNIT PRICE	EXTENDED AMOUNT
743359 SOD HYPO 12.5% LIQUICHLOR 1.0000 GL GL UNV LBL BLEACHTC NSF LIQ BULK	N	5000.00 GL	4504.00 GL	4504.0000 0.7500 GL	3378.00
698323 TRANSPORTATION SURCHG 1.0000 EA EA DIRECT BULK SPCL CHG **** **** NA	N	1.00 EA	1.00 EA	1.0000 0.0000 EA	0.00

TRANSPORTATION SURCHARGE APPLIES TO EACH DELIVERY
SURCHARGE RATES ARE INDEXED WITH TRANSPORTATION MARKET
CONDITIONS AND SUBJECT TO CHANGE.

PO 2017-00000195

SAFETY DATA SHEETS (SDS) ARE AVAILABLE ELECTRONICALLY
FOR UNIVAR DISTRIBUTED PRODUCTS AT
HTTPS://SDS.UNIVAR.COM.
SAFETY DATA SHEETS FOR PROPRIETARY PRODUCTS ARE
AVAILABLE AT THE UNIVAR CUSTOMER
SELF-SERVICE PORTAL, HTTPS://CUSTOMER.UNIVAR.COM.
IN THE EVENT OF AN INCIDENT, ALL SDSS MAY BE ACCESSED
BY EMERGENCY RESPONDERS BY CALLING 800-289-8892.

BULK CHEMICAL INVOICE

We deal with dangerous and volatile chemicals. It is imperative that you wear the proper personal protective equipment (PPE). Some standard items of PPE used for receiving chemical loads would be: a face shield, goggles, chemical-grade gloves, jacket, and pants.

Once the driver is onsite and is parked in the proper offloading position, you need to ensure that the truck gets connected to the right fill riser. The riser is where the truck's hose is connected to fill your bulk tank. Bulk chemical risers are normally locked. Unlock the correct fill riser. Have the driver connect his hose and open the fill valve. I do not allow my Operators to connect the driver's hose or open the fill valve. Figure 30-5 shows a chemical unloading station. Notice the grating beneath the fill connections; this is another spill containment. This area is external to the tanks but between the delivery truck and the fill connection.

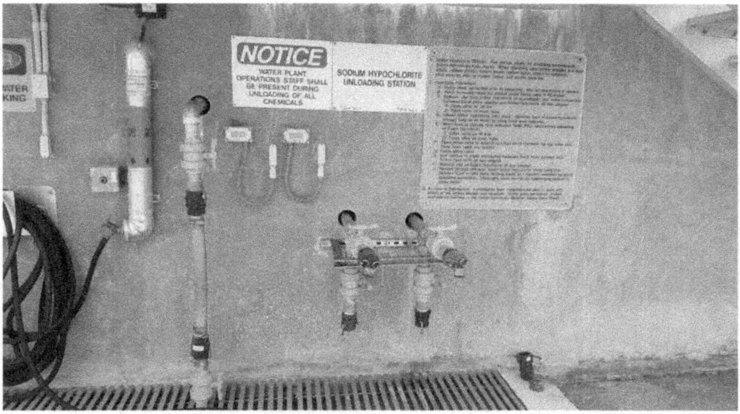

CHEMICAL UNLOADING STATION

Many plants test the bulk chemicals they receive. We won't address that here but remain focused on the receipt. When the driver begins to offload the chemical, verify that there are no leaks. Usually a drip bucket is placed under the hose connection to the tanker and another at the connection to the fill riser. If there are any leaks, I would have the driver stop offloading and fix the leak. When a leak happens, a situation may arise where the driver says the leak is the fault of your equipment. In these cases you might want to get maintenance and/or your supervisor involved. In any event, I wouldn't get into a confrontation with the driver.

When the driver begins offloading, you can tell because the hose will initially shake and then it will calm down. You can feel the chemical as it travels through the hose. Verify that you are filling the correct tank. You can verify this on the local-level display indicator and from the SCADA. I would check at both places, ensuring the correct tank is rising.

Some plants require the Operator to stay onsite during the entire load from start to finish. Follow your plant's Standard Operating Procedures. Continue to monitor the tank's level as it is being filled. There's never an excuse for overflowing a tank. Figure 30-6 shows digital tank level indicators, positioned locally at the fill risers. Figure 30-7 is a photo of a walled tank spill containment area. The containment pit has built-in pumps that facilitate removal of any spilled chemicals and any accumulated rainwater.

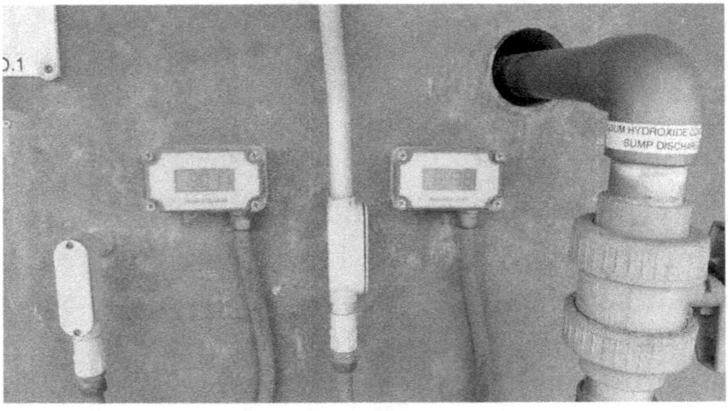

LOCAL TANK LEVEL INDICATORS

SPILL CONTAINMENT AREA

Once complete, the driver should close the fill valve and disconnect the hose. There shouldn't be a big flush of chemical spilling from the hose after the driver disconnects. If so, your plant's policy for chemical spills should be followed.

With the offload complete and all the appropriate paperwork in hand, lock the riser, check the loading area for debris and cleanliness, and deliver the paperwork as required. Don't forget that receipt of a bulk chemical load is a major event and should be logged in the Shift Report and other places as your plant requires.

The methods and practices discussed above are tried-and-true examples of good engineering practice designed to avoid inadvertent mixing of volatile chemicals, overflow of tanks, and taking on the wrong chemicals.

Real World 101:

The story below, as with all real-world information, actually happened. These types of occurrences are not uncommon. I call this one "The Incident."

On a normal weekday, two veteran Class 1 Operators were on shift. These Operators had more than 45 years of experience between them. This was a routine day and nothing special or out of the ordinary was happening. A load of 25 percent caustic was scheduled to arrive, and both Operators were aware of this.

The caustic load arrived in the morning as expected. The plant had two caustic tanks. Tank one was 89 percent full and was online. The offline tank was 20 percent full and had more than enough room to receive the entire caustic load, about 4,500 gallons. The Operators concurred that the offline tank was the one that should be filled.

The most senior Operator went out to receive the load. So far all was well. The driver had the appropriate paperwork and everything was in order.

The Operator unlocked the fill riser for the online tank instead of the offline tank. The driver made the connection. There were no leaks as the driver began to offload the caustic into the online tank. Remember, the online tank was already 89 percent full. Operators do not count on drivers to make sure their loads will fit or that the chemical is even going into the right tank for that matter.

The Operator receiving the load left the area and failed to check that the proper tank was rising. In a few minutes, the online tank reached 90 percent and the "high-level" alarm went off in the control room from SCADA and locally at the tank. Both the driver and the Operator in the control room acknowledged the alarm and went about their business. Wait for it: it gets better. Meanwhile the Operator who received the load initially was running about his daily tasks and still hadn't bothered to check on the progress of his caustic delivery.

Due to some very thoughtful engineering, these chemical bulk tanks were equipped with not one but two level alarms. The "high" level was set at 90 percent full and another separate and distinct "high high" level was set at 95 percent.

Approximately ten minutes after the "high" alarm went off, the "high high" alarm went off. Again, the control room Operator acknowledged the alarm and did absolutely nothing, for reasons no one can explain, not even the Operator in question. The driver was totally oblivious to the "high high" alarm, as he was sitting in his cab reviewing his Facebook page.

The tank was located inside a building and was surrounded by a concrete wall. The concrete wall comprised the containment area. The caustic continued to flow into the online tank and began to overflow into the containment area. The containment area was adjacent to several offices.

As the so-called containment area began to fill with caustic, the chemical breached (got outside of) the containment area. It began to penetrate the wall and enter an adjacent office. A chemist occupying the office noticed a white haze forming in the office and a pungent odor in the air. The chemist evacuated, warning nearby co-workers and notifying maintenance of the situation. Maintenance quickly assessed the problem, stopping the caustic transfer. Chemical spill protocols were quickly implemented, and as a result no personnel were hurt. There was also no need to evacuate the residents. However, as the damage assessment progressed, the repair cost careened into the tens of thousands of dollars.

So, we see that the Operators failed to act upon at least four checks and warnings, any one of which would have prevented this disaster:

1. The receiving Operator never checked to verify that the right fill riser was unlocked.
2. Both Operators failed to verify that the correct tank was rising.
3. The Control Room Operator acknowledged, then disregarded, the "high" alarm.
4. The Control Room Operator acknowledged, then disregarded, the "high high" alarm.

Figure 30-9 is a helpful checklist for receiving bulk chemical loads.

FIGURE 30-9

CHEMICAL LOAD CHECKLIST

1. Chemical .... ____________________

2. Split load.. YES _____ NO _____

3. Driver has the appropriate paperwork .... YES _____ NO _____

4. Tank designated for fill .... ____________________

5. Correct fill riser unlocked .... ____________________

6. Connection made and off load started (time) .... ____________________

7. Leaks .... YES _____ NO _____

8. Leaks corrected (if applicable) .... YES _____ NO _____

9. Correct tank filling verified (SCADA or locally)........... YES _____ NO _____

10. Offload complete, hose disconnected, area clean.....YES _____ NO _____

11. Fill riser locked ....YES _____ NO _____

12. Log entries made (all appropriate) .... YES _____ NO _____

BULK CHEMICAL RECEIPT CHECKLIST

Figure 30-10 shows a photo of a modern-day tank farm.

TANK FARM

Chapter 31

Sampling

Water treatment plant Operators are required to take many samples. Some are state and federally mandated. Others are prescribed by the plant manager or other supervisors. The best plants seem to sample beyond what is legally required and seem to have few or no violations.

In this chapter, we will discuss common samples that most plants draw and test. Operators must draw and test various types of samples. You need to know how to run the various tests, and understand and interpret what the results mean.

The most common samples and tests are turbidity, free and total chlorine, free ammonia, mono-chloramine, UV-254, permanganate, and alkalinity.

Turbidity:

Turbidity samples are very important because turbidity can tell you much about the status of your treatment. Not to mention, many of these samples are mandated by federal and state agencies. Most required samples are identified not only by location but by time interval and maximum turbidity. Some of the most likely turbidity samples are raw water, clarifier, sedimentation basin, filter, and finished water samples.

Free and Total Chlorine:

If your plant is using chlorine, no matter where or how it is being employed, it will be sampled and the results will be recorded. Disinfection is the most common use for chlorine, but it's also used as a pretreatment oxidizer.

Chlorine samples are normally taken at each application point to monitor the residual. These samples are taken at two- to four-hour intervals.

Regulatory agencies like the Department of Health Office of Drinking Water will often set minimum chlorine residual levels to be maintained throughout or at certain points in the treatment process.

Free Ammonia and Mono-Chloramine:

Plants that use mono-chloramine disinfection may test for mono-chloramine and free ammonia residuals. These two residuals are usually monitored by an online analyzer. Operators don't routinely sample and test for these residuals. However, there are occasions where you may be required to test for mono-chloramine and free ammonia. Instances requiring the testing for these residuals can occur if the mono-chloramine analyzer fails, gives suspect readings, or if you want to simply compare test results against the analyzer. In the case of total failure, Operators are usually required to run a bench test at prescribed intervals. This test is available as a single protocol that yields both results.

Mono-chloramine residuals between 3.0 and 4.0 mg/l are typical. For free ammonia, I instruct my Operators to keep the residual between 0.100 and 0.003 mg/l. Keep in mind that each plant will set their desired operating ranges.

Permanganate:

Permanganate is an oxidizer used mostly for iron and manganese removal. When it's used, the plant will set a target residual. Samples are normally taken just downstream of the application point. Then bench tests are run to ascertain the permanganate residual in the raw water. A common periodicity for this type of sample is once per shift.

Alkalinity:

Samples for alkalinity are very important, as you must have some alkalinity to achieve coagulation. Alkalinity samples are generally taken on raw, applied (water that has a primary coagulant added), and finished water. Operators are most concerned with the alkalinity of applied water. Applied alkalinity is critical and may be tested as frequently as every two hours.

UV-254:

The UV-254 test is used to measure organic matter in water. It utilizes ultraviolet (UV) absorbance. The test is also referred to as the Spectral Absorption Coefficient (SAC). The UV-254 nanometer (nm) wavelength is used to detect levels of organic matter in water.

When you run the UV-254 test on a raw water sample and a finished water sample you can calculate the percent removal of organics. Trending the percent removal can give you a good idea of the effectiveness of your treatment process. Plants that run this test usually run it once per day, logging the results for review and analysis.

Chapter 32

Reporting Violations

From an Operator's standpoint, reporting Violation'-s will always begin and most likely end with informing the Operations Supervisor or the Plant Manager. You should, however, be aware of the process and the possible ramifications that can occur when the plant is in violation.

The Operator should be concerned with two main areas in which violations are most likely to happen. That is in the areas of turbidity and disinfectant level. These areas are also the most preventable. Turbidity levels and disinfectant residuals are not the only areas in which you can find yourself in violation. For example, many plants fall under the requirement for online monitoring of certain parameters. Should the online monitoring system be disabled or go offline, there are actions that need to be taken. You may find your plant in violation even though the monitored parameter never got out of compliance. The violation resulted because the parameter was not being monitored.

Regulators understand that equipment failures happen. Specific procedures have been put in place for when equipment breakdown occurs. Most likely, manual sampling at specified intervals is implemented. These samples are read on bench test equipment and recorded by hand.

All states have their own mechanisms and procedures for reporting violations. Let's first understand where these regulations come from. Regulations and rules governing the standards to be maintained by water treatment plants are set by federal agencies, state authorities, and nationally recognized associations.

The Environmental Protection Agency (EPA), State Health Departments, and the American Water Works Association (AWWA) come quickly to mind. All these different groups have input on the regulations that we as Operators are required to uphold and enforce.

Most of these rules and regulations stem from the federal government, so the directives that we follow are similar throughout the United States. States, local agencies, national organizations, and especially treatment plants themselves will often operate under stricter guidance to avert plant violations and bolster public safety.

It is incumbent upon us as Operators to be subject-matter experts regarding our own plants' operating procedures as well as state and federal guidelines. For example, I know plants that have standing orders to remove online filters from service in the event that online turbidity rises above 0.2 NTU or spikes to 0.3 NTU or greater. This practice is more stringent than that required by the regulations for online filter turbidity.

When a plant find itself in violation it can trigger a host of undesirable consequences. One such consequence can come in the form of a "Consent Order." Consent Orders are agreements between two or more parties like the municipal water plant and the Department of Health. These orders and how they are implemented can vary by jurisdiction. They generally have the teeth of a court order and can be enforced by courts should any party fail to fulfill the obligations of the consent order. Other unpleasantries that can happen are: public and boiled-water notices, fines, loss of employment, and even jail time.

Many plants that are ahead of the power curve have flowcharts that help to assist Operators in recognizing what constitute violations and the actions they must take. The Environmental Protection Agency (EPA) has an intricate flowchart that gives guidance to be followed regarding individual filter requirements and reporting. The flowchart can be found in the EPA Guidance Manual Turbidity Provisions, Chapter 2 Turbidity Requirements: IESWTR, Figure 2-5, pp. 2–10.[1] See Figure 32-1.

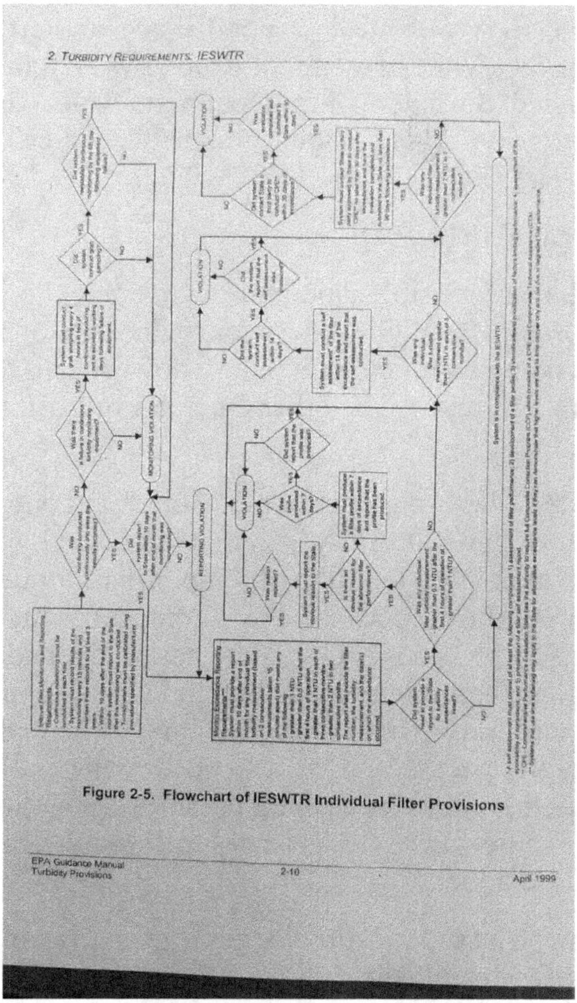

EPA TURBIDITY FLOWCHART

To stop violations and take appropriate actions you must take a vested interest in knowing what constitutes a violation. Familiarize yourself with the various rules and regulations that govern water treatment. The Plant Manager shouldn't have to point out that a violation has happened because you didn't realize a violation had occurred.

The EPA website is a good place to start. Review the Surface Water Treatment Rule (SWTR), the Interim Enhanced Surface Water Treatment Rule (IESWTR), and the Long Term 1 Enhanced Surface Water Treatment Rules (LT1ESWTR and LT2SWTR).

Real World 101:

The Responsible Operator in Charge, who is a Class 1 Operator (the highest license for the state of Virginia), is out doing chemical pump drawdowns. This task takes about 45 minutes. The other Operator on shift is a Class 4 Operator.

The Class 4 Operator is in the control room monitoring the SCADA system as directed by the Class 1 Operator. With ten filters online, filter #3 goes into an alarm condition because the filtered effluent turbidity has risen to 1.05 NTU. This has happened because the filter had high head loss and was approaching breakthrough. Breakthrough, or more accurately "Turbidity Breakthrough," can occur when the head loss gets so high that the trapped organics in the filter break through the filter and get passed through the effluent side. Not good.

The Class 4 acknowledges the alarm and makes note of it to tell the Class 1 what has occurred when the Operator gets back. The Class 4 doesn't realize that this high turbidity situation if left unattended can put the plant in violation.

At 2:45 P.M. the SCADA records the turbidity at 1.07 NTU. Fifteen minutes later at 3:00 P.M. the SCADA records the turbidity at 1.08 NTU. Now the plant is in violation.

The daily turbidity report is printed at midnight and shows the recorded violation. Unfortunately, the violation is found by the Assistant Water Production Manager the following morning and not the Operators. The violation occurs because online turbidities of greater than 0.300 NTU were recorded in two consecutive 15-minute intervals for an individual online filter.

The Assistant Water Production Manager immediately brought the violation to the Plant Manager's attention. The Plant Manager is required to inform the Department of Health Office of Drinking Water. Explanations were given to the Health Department in the form of a painstakingly detailed letter. The Plant Manager with input from the Water Quality Manager writes the letter. The letter must be reviewed and signed by the Director of Public Utilities. This all happens after an excruciatingly long, face-to-face conversation with the Director, explaining why and how the plant incurred a violation. In this case no fines, sanctions, or consent orders were levied. The Health Department recommended special training for all Operators, and by recommended, I mean ordered. The Class 1 Operator was sent home without pay for three days and the Operator in training was given a written reprimand held on file for one year pending no other incidents.

A Class 4 Operator should know what situations would put the plant in violation and at a minimum notify the Responsible Operator in Charge immediately when the first turbidity reading above 0.300 NTU occurred. Taking the filter offline would have averted a major plant violation.

References:

1. EPA Guidance Manual Turbidity Provisions, Chapter 2 Turbidity Requirements: IESWTR, Fig 2-5, pp. 2–10, 1996.

SECTION 7

Troubleshooting

Chapter 33

Troubleshooting (It's a Process)

Troubleshooting is an ancient art that is being lost among Operators. Super Operators are usually very good troubleshooters. You may be asking why this is an important skill to have in your arsenal. Simply put, it helps the maintenance technician by pointing him in the right direction when trying to figure out what's wrong. The ability to troubleshoot allows you to provide more information about the casualty or problem. It also helps you understand what immediate actions to take to correct a situation that could potentially put your plant in violation.

The Operator who writes a work order that says "pump stopped" hasn't done any troubleshooting and is leaving the maintenance technician in the dark. It is the maintenance technician's job to repair equipment and solve systemic problems, but the Operators are the most knowledgeable about how the systems and equipment operate. You're the one who can provide firsthand knowledge when casualties happen. You should be the subject-matter expert on how the equipment runs and when it's not functioning properly.

Troubleshooting a problem is a methodical process based on proven techniques that yield accurate and rapid results. The first step in troubleshooting is recognizing you have a problem. Knowing your plant and how it runs is paramount if you expect to be a thorough troubleshooter. For example, you should know that in automatic mode, when the high-service pump station discharge pressure drops below 50 psi, a second pump should come online. If you don't know this, when the pump doesn't come on, you won't see it as a problem.

The following is my step-by-step chart of how to troubleshoot any problem:

Step 1:

Recognize there is a problem.

(a) The level in the clear well is not changing.

Step 2:

Find out if the system or equipment has been recently worked on.

(a) You assume the shift and notice the level in your clear well is not changing, which is unusual for the time of day and mode of operation. Using your professional knowledge, you know that the level should in fact be dropping.

(b) By checking the shift report you see that an outside contractor was onsite and had calibrated the clear well's level transmitter. Your problem may not be related to, but most likely lies in, the contractor's work on the transmitter.

Step 3:

Get out of the chair.

(a) Many Operators languish in mediocrity because they refuse to get out of the control room chair and look about the plant. In other words, you should continue troubleshooting.

Especially if there are no obvious signs on SCADA, such as a negative reading that may indicate power loss to the sensor or equipment.

(b) It is a must that you go to the failed gear and look for obvious and not-so-obvious problems. Your casualty could simply be a cord that has been unplugged, a frayed or broken wire, a switch placed in the wrong position.

(c) Getting out of the chair and investigating the problem helps you write a comprehensive work order and maybe even correct the problem without having to write a work order.

Note: A complete and well-written work order also helps the maintenance technician a great deal. It also strengthens your relationship with the maintenance team. The truth is that Operators who are regarded as knowledgeable and helpful to the team effort get better results. Maintenance and all other entities (lab technicians, contractors, etc.) are overwhelmingly willing to work with Operators they see as engaged and concerned about the plant.

Step 4:

Ascertain what would cause the problem that you are seeing.

(a) Intimate knowledge about your plant and how it works helps especially when the problem is not obvious from initial investigation. You will need to think hard about what things could possibly cause this failure. Plug along until you solve the problem or get some idea as to which direction the problem lies. Sometimes you can't always solve the exact problem. At least you have held a reasonable investigation and not simply said "the pump won't start."

(b) Some possible causes will be ruled out immediately because it will be obvious that they could not have caused that problem.

The ability to investigate and troubleshoot problems is a priceless ability that will help you immensely as an Operator. Proficiency in troubleshooting provides for quick restoration of upsets in the treatment process.

The above four steps are simple, yet they will yield great results when employed. The key lies in knowing how your plant works.

Real World 101:

Case 1:

I routinely walk into the control room and pose hypothetical situations to the Operators. I call it "Paper Troubleshooting." These exercises get them used to thinking logically about problems. Many times, the exercises reveal misunderstandings about how the plant works. I have found that these "Paper Troubleshooting" sessions are excellent teaching tools.

I asked a trainee what could cause his applied pH to rapidly begin to fall. The system used caustic to automatically regulate pH, and alum was the primary coagulant.

The trainee responded by saying, "We could be overfeeding carbon."

I said, "Let's think about that. Carbon might have a very, very slight effect on pH but an overfeed would surely cause the pH to increase, not decrease. The reason it would possibly rise is because carbon is not acidic, it's more basic in nature. Also, you would most likely see some increase in turbidity if you were overfeeding carbon."

I said, "Look at the caustic screen on SCADA. We see that the caustic pump is being ordered to increase speed, because the system is trying to correct for the low applied pH. What if your caustic pump is failing?"

We continued our paper troubleshooting exercise and I said, "We should go and draw down the pump. Okay, we draw down the caustic pump and find that the caustic pump wasn't pumping at all. What should we do?"

My trainee replied, "We should switch pumps and write a work order."

"Excellent!" I told the trainee. The pump wasn't pumping because the roller had popped out of place. The failed caustic pump explained all the symptoms we were seeing in this scenario.

Case 2:

While monitoring SCADA, the #2 sedimentation basin high-level alarm starts going on and off intermittently. It goes high, then suddenly returns to normal.

Since you're regulating multiple filter flows based on this level reading, you need it to be accurate. An error in the basin level reading can cause flooding or drain your filter beds, both of which are undesirable.

You know that this sensor is a sonic level sensor and you know where it's located. You get up from the chair in the control room and go look at the sensor.

Once onsite, the problem shows itself and is easily corrected. Since the sensor is sonic, it requires a clear path from the sensor to the surface of the level it is trying to detect.

There is an overgrown bush near the sensor whose branches are blowing intermittently between the sensor and the surface of the basin. This would explain the exact behavior of the malfunction.

Tying the branches back fixed the problem, and they can be trimmed back properly later. Knowing the plant and how it operates made it an easy fix. Had the operator not known where the sensor was located, or that it was sonic and how sonic sensors work, it could have meant calling in maintenance for a very minor problem.

SECTION 8

Appendix A: Operator Qualification Standards

Appendix B

Operator Qualification Standards Defined

Training Operators is one of the most important and difficult tasks of Plant Managers, Superintendents, and Senior Operators. Training them to operate the plant uniformly is even more difficult. Appendix A is a system designed to help train Operators in a consistent and uniform manner. Appendix A also enables managers to document and certify that their Operators have been properly trained. Appendix A can be used as is or edited to fit your specifications. Managers designate which Operators can sign off the qualification standards. In some cases, oral boards and/or written examinations may be required for final certifications.

Appendix A has four major sections. They are Fundamentals (Series 100), Systems (Series 200), Tasks (Series 300), and Special Evolutions (Series 400). Each section contains line items that must be mastered through demonstration and then signed off as completed by designated personnel. Each series can be accomplished simultaneously. Descriptions of the major sections follow:

Fundamentals:

Series 100 – In this series Operators demonstrate knowledge of the fundamentals (basics) of conventional water treatment. These fundamentals lay the foundation required for them to develop into "Super Operators." The fundamentals section ensures the Operator knows basics such as coagulation, and can answer questions such as: What is floc? What must you have for coagulation to occur? Define turbidity, etc. As with all sections in Appendix A, once the Operator has satisfactorily demonstrated that they have retained the requisite information, sign off the line item as completed.

Systems:

Series 200 – Contains the qualification standards that cover the plant's systems—the integral parts of the plant such as individual chemical feed systems, raw flow, and the high-service pump systems. Trainees must demonstrate knowledge of equipment, their location, number, capacity, and operational modes. Systems ensures the Operator knows information necessary to properly run the plant, keep it in compliance, and deal with casualties.

Tasks:

Series 300 – The Tasks portion of the qualification standards includes physical demonstration and successful accomplishment of various routine actions required of Operators. Tasks include procedures like: chemical pump drawdowns, various water chemistry tests, and the use of various types of lab equipment. Management may require more than one successful completion to be fully qualified in a task.

Special Evolutions:

Series 400 – Special Evolutions covers the ability of trainees to demonstrate the successful accomplishment of nonroutine special evolutions. These evolutions include placing clarifiers or sedimentation basins in and out of service, switching rapid mix trains, and placing systems in unusual configurations such as isolating and bypassing a clear well for divers to inspect. Since these types of events do not occur with any great frequency, many plants use the walk-through demonstrations method to get the Operator closer to qualification. When possible, schedule these events when Operators requiring the certification are on shift.

Appendix C

Fundamentals

Series 100

In this series, Operators demonstrate knowledge of the fundamentals of conventional water treatment. These fundamentals lay the foundation required for them to develop the necessary skills to properly operate the plant.

The trainee should be able to satisfactorily answer the questions in the 100 series. Once completed, an authorized person should sign and date the question. Management may require an oral board or written examination as the final step to qualification of a section.

	Questions:	Signature / Date
100.1	What is coagulation?	______________
100.2	What is flocculation?	______________
100.3	What is floc?	______________
100.4	Raw surface water typically has what type of charge?	______________
100.5	What must be present in raw water to achieve coagulation?	______________
100.6	Define turbidity.	______________
100.7	What is disinfection?	______________

100.8	What primary coagulant does this plant use?	________
100.9	What is oxidation?	________
100.10	What is polymer used for?	________
100.11	What is chlorine used for? (Name two uses.)	________
100.12	What is carbon used for?	________
100.13	What is ammonia used for?	________
100.14	What is zinc orthophosphate used for?	________
100.15	What is potassium permanganate used for?	________
100.16	Define absorption.	________
100.17	Define adsorption.	________
100.18	What are mono-chloramines?	________
100.19	What are coagulants used for?	________
100.20	What are oxidizers used for?	________
100.21	Define disinfection.	________

100.22 What are filters used for? ______________

100.23 Define effluent. ______________

100.24 Define influent. ______________

100.25 Define low-service pumps. ______________

100.26 Define high-service pumps. ______________

100.27 What are clarifiers? ______________

100.28 What are sedimentation basins? ______________

100.29 What is caustic soda used for? ______________

100.30 Define pressure. ______________

100.31 What does psi stand for? ______________

100.32 Define suction. ______________

100.33 What is corrosion? ______________

100.34 Define head loss. ______________

100.35 How many pounds per gallon of water? ______________

100.36 How many gallons in a cubic foot? ______________

100.37 What does mg/l stand for? ______________

100.38 Define positive displacement pump. ______________

100.39 Define centrifugal pump. ______________

100.40 What is a flow meter used for? ______________

100.41 What is a strainer? ______________

100.42 Define differential pressure. ______________

100.43 One psi of water equals how many feet of water? ______________

100.44 Define meniscus. ______________

****
****

Oral board required
(yes ____ no ____) ____________________

Written examination required
(yes _____ no _____) ____________________

Final Certification: ____________________

Signature / Date

Appendix D

Systems

Series 200

Series 200 covers the plant's systems. Trainees must demonstrate knowledge of equipment, purpose, basic operation, automated functions, number, and capacity. Systems ensure the Operator knows the information necessary to properly run the plant, keep it in compliance, and deal with casualties.

RAW WATER SOURCES:

200.1 Surface water sources (list all):

____

____

Signature / Date

200.2 Ground water sources (list all):

____

____

Signature / Date

PRETREATMENT:

200.3 Type of pretreatment:

____

200.4 Chemicals and/or method used: ___________

____

200.5

Chemical # of Pumps Capacity Mode(s) of Operation Cross-Connectable (Y/N)

_______ _____ ______ _____ ______

____

Signature / Date

SEDIMENTATION / CLARIFICATION:

200.6 Type: ______________________________

Number of Units: __________________________

Volume / Capacity: _________________________

Sludge removal process: _____________________

____

____

Signature / Date

CHEMICAL FEED SYSTEMS:

200.7 Bulk storage:

Chemical type (% strength if applicable)	Number of tanks and tank volume

Signature / Date

200.8 Chemical Feed Pumps:

Chemical	# of Pumps	Capacity	Mode(s) of Operation	Cross-Connectable (Y/N)

Signature / Date

FILTRATION SYSTEM:

200.9 Type: ______________________________

200.10 Total number: ______________________

200.11 Filter size(s): _____________________

____

200.12 Are filters grouped? ______ If so, describe: ___________________________________

____

200.13 When does the high head loss alarm trigger?

____

200.14 How many hours can a filter stay online before it must be washed? ______________________

____
Signature / Date

LOW LIFT PUMP SYSTEM:

200.15

# of Pumps	Type	Capacity	Mode(s) of Operation
________	_____	_________	______________

200.16 Describe pump logics (if applicable): ______________________________

____

____

*Pump logics are associated with the automated functions such as shutdown or start, based on programmed parameters. For example, in auto-mode the standby pump comes on if the lead pump cannot maintain a certain level for a certain period.

____
Signature / Date

FINISHED WATER STORAGE:

200.17 # of Clear Wells: ________ Capacity: ____

____
Signature / Date

HIGH-SERVICE PUMP SYSTEM:

200.18

# of Pumps	Type	Capacity	Mode(s) of Operation
________	_____	_________	______________

200.19 Describe pump logics (if applicable): ______________________________
____
____

____
Signature / Date

DISTRIBUTION TANKS:

200.20

Tank Name	Capacity	Level When Full	Minimum Operating Level
________	________	________	________
________	________	________	________
________	________	________	________
________	________	________	________
________	________	________	________
________	________	________	________
________	________	________	________
________	________	________	________

____

Signature / Date

"YOU ARE A DROP OF WATER. TAKE ME FROM THE SOURCE TO THE CLEAR WELL."

200.21

Describe in detail each step a drop of raw water goes through once it leaves the source (raw) water and starts on its way through the treatment process, ending up in the clear well. No detail is too small. All the major points of interest, such as when chemicals are added, should be noted.

____

Signature / Date

****
****

Oral board required
(yes ___ no ____)

Signature / Date

Written examination required
(yes ___ no ____)

Signature / Date

Final Certification:

Signature / Date

Appendix E

Tasks

Series 300

Tasks: The Tasks or Series 300 concerns qualification standards establishing the Operator's ability to accomplish routine tasks. Tasks include procedures such as chemical pump drawdowns, various water chemistry tests, and the use of various types of lab equipment. Management may require more than one successful completion to be fully qualified in a task.

ROUTINE SAMPLING:

300.1 Successfully demonstrate how to properly draw the following samples:

a. Clarifier / Sedimentation Basin Turbidity sample	______________________ Signature / Date
b. Free Chlorine sample	______________________ Signature / Date
c. Total Chlorine sample	______________________ Signature / Date
d. Finished Water Turbidity sample	______________________ Signature / Date
e. Permanganate residual sample	______________________ Signature / Date

f. Mono-chloramine residual sample ______________________ Signature / Date

g. Free Ammonia residual sample ______________________ Signature / Date

h. Applied pH sample ______________________ Signature / Date

i. Fluoride (well) residual sample ______________________ Signature / Date

WATER CHEMISTRY TESTS:

Successfully demonstrate the proper testing procedure and use of equipment for the following:

300.2 Free Chlorine residual ______________________ Signature / Date

300.3 Total Chlorine residual ______________________ Signature / Date

300.4 Alkalinity ______________________ Signature / Date

300.5 Mono-Chloramine residual ______________________ Signature / Date

300.6 Free Ammonia residual ______________________ Signature / Date

300.7 True Color ______________________ Signature / Date

300.8 pH meter ______________________ Signature / Date

300.9 ISE (Ion Selective Electrode) meter ____________________ Signature / Date

300.10 Permanganate residual ____________________ Signature / Date

300.11 Titration procedure ____________________ Signature / Date

300.12 Bench Top Turbidimeter ____________________ Signature / Date

300.13 Laboratory Spectrophotometer (Example DR 3900) ____________________ Signature / Date

300.14 Auto Burette ____________________ Signature / Date

300.15 Manual Burette ____________________ Signature / Date

ROUTINE TASKS:

Exhibit proficiency in accomplishing the following common tasks:

300.16 Chemical Pump Drawdowns ____________________ Signature / Date

300.17 Calculate feed rates in mg/l using drawdown and other data ____________________ Signature / Date

300.18 Maintain Chemical Drawdown Sheet ____________________ Signature / Date

300.19	Rotate Chemical Feed Pumps	Signature / Date
300.20	Maintain the Operator Bench Sheet	Signature / Date
300.21	Maintain the Shift Report	Signature / Date
300.22	Perform a Filter Backwash	Signature / Date
300.23	Maintain Filter Backwash Log	Signature / Date

SCADA SYSTEM:

Demonstrate mastery of the following tasks regarding the SCADA system:

300.24	Describe key points and functions of each SCADA screen.	Signature / Date
300.25	Demonstrate use of the SCADA trending function.	Signature / Date
300.26	Operate equipment from SCADA.	Signature / Date

****

Oral board required (yes ____ no ____)	______________________ Signature / Date
Written examination required (yes _____ no ____)	______________________ Signature / Date
Final Certification:	______________________ Signature / Date

Appendix F

Special Evolutions

Series 400

Special Events:

Series 400 – Special Evolutions covers the ability of trainees to demonstrate the successful accomplishment of nonroutine special evolutions. These evolutions include placing clarifiers or sedimentation basins in and out of service, switching rapid mix trains, and placing systems in unusual configurations such as isolating and bypassing a clear well for divers to inspect. Since these types of events do not occur with any great frequency, many plants use the walk-through demonstrations method to get the Operator closer to qualification. When possible, schedule these events when Operators requiring the certification are on shift.

SPECIAL EVOLUTIONS:

400.1 Successfully demonstrate how to properly perform the following special evolutions:

a. Place a Clarifier/Sedimentation Basin into service ______________________ Signature / Date

b. Take a Clarifier/Sedimentation Basin out of service ________________ Signature / Date

c. Perform a Filter Drawdown ________________ Signature / Date

d. Switch Rapid Mix Trains ________________ Signature / Date

e. Cross-connect chemical feed pumps (ex. alum pump 3 feeding train 1) ________________ Signature / Date

f. Operate the primary coagulant system manually and regulate applied pH ________________ Signature / Date

g. Manually regulate the ammonia feed ________________ Signature / Date

h. Bypass a clear well for inspection ________________ Signature / Date

****
****

Oral board required (yes ____ no ____)	______________________ Signature / Date
Written examination required (yes _____ no ____)	______________________ Signature / Date
Final Certification:	______________________ Signature / Date

www.ingramcontent.com/pod-product-compliance
Lightning Source LLC
LaVergne TN
LVHW052319200926
848481LV00002B/609
* 9 7 8 0 8 2 0 6 0 4 1 3 8 *